VOLUME 669

THE ANNALS

of The American Academy of Political
and Social Science

*The New Big Science: Linking Data to
Understand People in Context*

Special Editors:
SANDRA L. HOFFERTH
University of Maryland
EMILIO F. MORAN
Michigan State University

Los Angeles | London | New Delhi
Singapore | Washington DC | Melbourne

The American Academy of Political and Social Science

202 S. 36th Street, Annenberg School for Communication, University of Pennsylvania,
Philadelphia, PA 19104-3806; (215) 746-6500; (215) 573-2667 (fax); www.aapss.org

Board of Directors
KENNETH PREWITT, *President*
JANICE MADDEN, *Chair*

ANDREW J. CHERLIN
SHELDON DANZIGER
GREG DUNCAN
JACOB HACKER
ROBERT HAUSER

JAMES JACKSON
ARTHUR LUPIA
REBECCA MAYNARD
MARY ANN MEYERS
DOROTHY ROBERTS
THEDA SKOCPOL

Editors, THE ANNALS
THOMAS A. KECSKEMETHY, *Executive Editor*
EMILY W. BABSON, *Managing Editor*
PHYLLIS KANISS, *Editor Emerita (deceased)*   RICHARD D. LAMBERT, *Editor Emeritus*

Editorial Advisory Board

MAHZARIN BANAJI, *Harvard University*
FRANCINE BLAU, *Cornell University*
FELTON EARLS, *Harvard University*
PAULA ENGLAND, *New York University*
LEE EPSTEIN, *Washington University, St. Louis*
ROBERT GREENSTEIN, *Center on Budget and Policy Priorities*

ROBERT KEOHANE, *Princeton University*
DOUGLAS S. MASSEY, *Princeton University*
SEAN REARDON, *Stanford University*
ROGERS SMITH, *University of Pennsylvania*
THOMAS SUGRUE, *University of Pennsylvania*

Origin and Purpose. The Academy was organized December 14, 1889, to promote the progress of political and social science, especially through publications and meetings. The Academy does not take sides in controverted questions, but seeks to gather and present reliable information to assist the public in forming an intelligent and accurate judgment.

Meetings. The Academy occasionally holds a meeting in the spring extending over two days.

Publications. THE ANNALS of The American Academy of Political and Social Science is the bimonthly publication of the Academy. Each issue contains articles on some prominent social or political problem, written at the invitation of the editors. These volumes constitute important reference works on the topics with which they deal, and they are extensively cited by authorities throughout the United States and abroad.

Subscriptions. THE ANNALS of The American Academy of Political and Social Science (ISSN 0002-7162) (J295) is published bimonthly—in January, March, May, July, September, and November—by SAGE Publications, 2455 Teller Road, Thousand Oaks, CA 91320. Periodicals postage paid at Thousand Oaks, California, and at additional mailing offices. POSTMASTER: Send address changes to The Annals of The American Academy of Political and Social Science, c/o SAGE Publications, 2455 Teller Road, Thousand Oaks, CA 91320. Institutions may subscribe to THE ANNALS at the annual rate: $1070 (clothbound, $1209). Individuals may subscribe to the ANNALS at the annual rate: $122 (clothbound, $180). Single issues of THE ANNALS may be obtained by individuals for $38 each (clothbound, $52). Single issues of THE ANNALS have proven to be excellent supplementary texts for classroom use. Direct inquiries regarding adoptions to THE ANNALS c/o SAGE Publications (address below).

Copyright © 2017 by The American Academy of Political and Social Science. All rights reserved. No portion of the contents may be reproduced in any form without written permission from the publisher.

All correspondence concerning membership in the Academy, dues renewals, inquiries about membership status, and/or purchase of single issues of THE ANNALS should be sent to THE ANNALS c/o SAGE Publications, 2455 Teller Road, Thousand Oaks, CA 91320. Telephone: (800) 818-SAGE (7243) and (805) 499-0721; Fax/Order line: (805) 375-1700; e-mail: journals@sagepub.com. *Please note that orders under $30 must be prepaid.* For all customers outside the Americas, please visit http://www.sagepub.co.uk/customerCare.nav for information.

Printed on acid-free paper

THE ANNALS
© 2017 by The American Academy of Political and Social Science

All rights reserved. No part of this volume may be reproduced or utilized in any form or by any means, electronic or mechanical, including photocopying, recording, or by any information storage and retrieval system, without permission in writing from the publisher. All inquiries for reproduction or permission should be sent to SAGE Publications, 2455 Teller Road, Thousand Oaks, CA 91320.

Editorial Office: 202 S. 36th Street, Philadelphia, PA 19104-3806
For information about individual and institutional subscriptions address:
SAGE Publications
2455 Teller Road
Thousand Oaks, CA 91320

For SAGE Publications: Peter Geraghty (Production) and Mimi Nguyen (Marketing)

From India and South Asia, write to:	From Europe, the Middle East, and Africa, write to:
SAGE PUBLICATIONS INDIA Pvt Ltd	SAGE PUBLICATIONS LTD
B-42 Panchsheel Enclave, P.O. Box 4109	1 Oliver's Yard, 55 City Road
New Delhi 110 017	London EC1Y 1SP
INDIA	UNITED KINGDOM

International Standard Serial Number ISSN 0002-7162
ISBN 978-1-5063-8446-7 (Vol. 669, 2017) paper
ISBN 978-1-5063-8445-0 (Vol. 669, 2017) cloth
Manufactured in the United States of America. First printing, January 2017

Please visit http://ann.sagepub.com and under the "More about this journal" menu on the right-hand side, click on the Abstracting/Indexing link to view a full list of databases in which this journal is indexed.

Information about membership rates, institutional subscriptions, and back issue prices may be found on the facing page.

Advertising. Current rates and specifications may be obtained by writing to The Annals Advertising and Promotion Manager at the Thousand Oaks office (address above). Acceptance of advertising in this journal in no way implies endorsement of the advertised product or service by SAGE or the journal's affiliated society(ies) or the journal editor(s). No endorsement is intended or implied. SAGE reserves the right to reject any advertising it deems as inappropriate for this journal.

Claims. Claims for undelivered copies must be made no later than six months following month of publication. The publisher will supply replacement issues when losses have been sustained in transit and when the reserve stock will permit.

Change of Address. Six weeks' advance notice must be given when notifying of change of address. Please send the old address label along with the new address to the SAGE office address above to ensure proper identification. Please specify the name of the journal.

THE ANNALS
OF THE AMERICAN ACADEMY OF POLITICAL AND SOCIAL SCIENCE

Volume 669   January 2017

IN THIS ISSUE:

The New Big Science: Linking Data to Understand People in Context

Special Editors: SANDRA L. HOFFERTH and EMILIO F. MORAN

Introduction: History and Motivation .... *Sandra L. Hofferth, Emilio F. Moran, Barbara Entwisle, J. Lawrence Aber, Henry E. Brady, Dalton Conley, Susan L. Cutter, Catherine C. Eckel, Darrick Hamilton, and Klaus Hubacek*   6

The Promise, Practicalities, and Perils of Virtually Auditing Neighborhoods Using Google Street View............. *Michael D. M. Bader, Stephen J. Mooney, Blake Bennett, and Andrew G. Rundle*   18

Understanding Racial Differences in Exposure to Violent Areas: Integrating Survey, Smartphone, and Administrative Data Resources .... *Christopher R. Browning, Catherine A. Calder, Jodi L. Ford, Bethany Boettner, Anna L. Smith, and Dana Haynie*   41

Linking Federal Surveys with Administrative Data to Improve Research on Families .... *Amy O'Hara, Rachel M. Shattuck, and Robert M. Goerge*   63

Predicting Asthma Prevalence by Linking Social Media Data and Traditional Surveys..................... *Hongying Dai, Brian R. Lee, and Jianqiang Hao*   75

Correlates of Contraceptive Use and Health Facility Choice among Young Women in Malawi .... *Jean Digitale, Stephanie Psaki, Erica Soler-Hampejsek, and Barbara S. Mensch*   93

Understanding How Low–Socioeconomic Status Households Cope with Health Shocks: An Analysis of Multisector Linked Data .... *Tammy Leonard, Amy E. Hughes, and Sandi L. Pruitt*   125

Weather-Related Hazards and Population Change: A Study of
   Hurricanes and Tropical Storms in the
   United States, 1980–2012 . . . . *Elizabeth Fussell, Sara R. Curran,*
                                         Matthew D. Dunbar, Michael A. Babb,
                           Luanne Thompson, and Jacqueline Meijer-Irons 146

New Trends and Patterns in Western European Immigration
   to the United States: Linking European and American
   Databases . . . . *Elyakim Kislev* 168

Quilting a Time-Place Mosaic: Concluding Remarks . . . . *Barbara Entwisle,*
                                     Sandra L. Hofferth, and Emilio F. Moran 190

FORTHCOMING

Regulatory Intermediaries in the Age of Governance
Special Editors: KENNETH W. ABBOTT, DAVID LEVI-FAUR,
and DUNCAN SNIDAL

Student Debt: Effects, Opportunity and Federal Policy
Special Editors: LAURA PERNA and NICK HILLMAN

Introduction: History and Motivation

By
SANDRA L. HOFFERTH,
EMILIO F. MORAN,
BARBARA ENTWISLE,
J. LAWRENCE ABER,
HENRY E. BRADY,
DALTON CONLEY,
SUSAN L. CUTTER,
CATHERINE C. ECKEL,
DARRICK HAMILTON,
and
KLAUS HUBACEK

Big data, that is, data that are byproducts of our lives rather than designed for research purposes, are the newest of the information highway innovations. One of the important challenges to social and behavioral science data collection, curation, and dissemination for the foreseeable future is to link diverse forms of data in a way that is cumulative, representative, meaningful, and accessible to a broad range of researchers. It is critical to explore the new questions these data can address and to develop new methods to address them, including linking persons and information about them and their environments across different data platforms while maintaining confidentiality and privacy. Linking a broad array of information—from administrative data (local and state and regional), to social media (Twitter, Facebook), to census and other surveys, to ethnographic data, and data from experiments such as randomized controlled trials—to address how humans and their communities make decisions is challenging. This issue was addressed by papers presented at a conference on New Data Linkages convened by the Social Observatories Coordinating Network in 2016; those articles are brought together in this volume.

Keywords: big data; social science; systems science; data analytics; knowledge generation; survey research; administrative data

The nation and the world are changing rapidly, yet scholars attempt to understand such changes with tools and infrastructure that

Sandra L. Hofferth is a professor emerita, School of Public Health, and a research professor, Maryland Population Research Center, University of Maryland. Her research focuses on Americans' use of time; economic disadvantage, parental behavior, and child health and development; fathers and fathering; and immigrant youth's transition to adulthood.

Emilio F. Moran is Hannah Distinguished Professor at Michigan State University. He is the author of eleven books, fifteen edited volumes, and more than 200 journal articles and book chapters. His research addresses how humans interact with the environment. He was elected to the U.S. National Academy of Sciences in 2010.

Correspondence: hofferth@umd.edu

DOI: 10.1177/0002716216682715

were developed more than 50 years ago. In the last decade, hurricanes and floods have devastated coastal areas of the United States; sudden migratory flows due to environmental, political, and economic events abroad have raised concerns about the assimilation of refugees and immigrants; and the increased financial burden of coping with unexpected health shocks has drawn attention to inequities in the health care system and exposed patterns of ecological, economic, and social vulnerabilities across the nation. Finally, a large and robust middle class—one of America's greatest achievements—has been steadily eroding, leading to anger at our social and political institutions.

Barbara Entwisle is Kenan Distinguished Professor of Sociology and a fellow of the Carolina Population Center at the University of North Carolina at Chapel Hill. Her research focuses broadly on the study of social, natural, and built environments and their consequences for a range of demographic and health outcomes.

J. Lawrence Aber is the Willner Family Professor of Psychology and Public Policy at NYU Steinhardt, and codirector of the international research center, Global TIES for Children. His basic research examines the influence of poverty and violence on the social, emotional, behavioral, cognitive, and academic development of children and youth.

Henry E. Brady is the dean of the Goldman School of Public Policy and Class of 1941 Monroe Deutsch Professor of Political Science and Public Policy, University of California, Berkeley. His research and writings address economic and political issues including electoral politics and political participation, social welfare policy, political polling, and the dynamics of public opinion.

Dalton Conley is the Henry Putnam University Professor in Sociology at Princeton University. His research focuses on how socioeconomic status and health are transmitted across generations and on the public policies affecting those processes. He studies sibling differences in socioeconomic success; racial inequalities; and how health and biology affect (and are affected by) social position.

Susan L. Cutter is a Carolina Distinguished Professor of Geography at the University of South Carolina and director of its Hazards and Vulnerability Research Institute. Her research focuses on vulnerability and resilience metrics especially as they relate to social inequalities in disaster response and recovery.

Catherine C. Eckel is the Sara & John Lindsey Professor of Economics and university distinguished professor at Texas A&M University, where she directs the Behavioral Economics and Policy Program. She uses experimental methods to investigate social and economic influences on behavior, including topics such as discrimination, charitable giving, risk-taking, and trust.

Darrick Hamilton is the director of the doctoral program in policy and an associate professor of economics and urban policy at The New School in New York. He is a stratification economist, whose work examines the causes, consequences, and remedies of racial and ethnic inequality including intersections of identity, racism, colorism, and socioeconomic outcomes.

Klaus Hubacek is an ecological economist and professor in the Department of Geographical Sciences at the University of Maryland, College Park. His research focuses on conceptualizing and modeling the interaction between human and environmental systems and modeling scenarios of future change.

NOTE: Funding for the Social Observatories Coordinating Network was provided by National Science Foundation grant SES1237498.

Scientific explanations of these phenomena have been fragmentary and discipline-bound, so our prescriptions for reversing these trends are piecemeal and there is a high degree of uncertainty as to their effectiveness. Moving forward will require new ways of thinking and new ways of accessing, curating, and analyzing the existing but not always accessible information. Social scientists generally work with survey, administrative, observational, and experimental data, data gathered for specific purposes. Social scientists' infrastructure and tools consist of planned data collection and well-established analytic methods. Survey methods permit generalizing questionnaire responses to known populations, observational methods add depth, and experimental methods facilitate causal inference. Linked with surveys, administrative data could expand our understanding of individual behavior but have remained largely inaccessible. Examples of large-scale survey data of the federal statistical system include data held and disseminated by the National Center for Health Statistics, the U.S. Census Bureau, the Bureau of Labor Statistics, and the Economic Research Services of the Department of Agriculture. However, big data—large, diverse, and heterogeneous datasets, often by-products generated from business and Internet transactions, email, social media, health care facilities, and various sensors and instruments—have produced large archives of data that are not organized in a way that can be easily analyzed by social scientists (Foster et al. 2017). Population characteristics are often unknown and inferential statistics inappropriate. However, these new data provide information, such as emotional responses, that may complement traditional types of data and are more immediately accessible. To make this situation even more challenging, new social phenomena, particularly social media (e.g., Twitter, LinkedIn, Facebook), have arisen that cannot be studied with traditional methods, but our academic programs do not prepare the new generation of social scientists to link new media with other kinds of social data. We need to improve our tools and the information that we generate to better serve business, government, and the social and economic needs of the population.

Developments in information technology offer an unprecedented opportunity to collect diverse data at fine-grained spatial and temporal scales, and present a remarkable chance to change the way social science is conducted and to greatly expand the questions that can be addressed. Today, the proliferation of new data coming from the Internet and social media requires new ways to collaborate across social science disciplines and to link social science with genetic, linguistic, medical, environmental, biological, and earth systems science. This is an opportune time to rethink the primary ways in which data are collected, gathered, coded, curated, documented, archived, and disseminated in the United States. This volume builds on a series of workshops sponsored by the National Science Foundation (NSF) in 2005, 2007, 2008, 2009, 2010, and 2011 that led to a consensus for the creation of a network of regional data centers that could, when fully developed, represent the entire U.S. population and its diverse regions. In 2012, the NSF began to support a set of researchers and academic faculty from across the United States, known as "the Social Observatories Coordinating Network (SOCN)," to take on this challenge.[1] In our several years of discussions, the network settled on a challenge for the future—to design a national network of regional data centers that could be coordinated through common objectives, sharing of protocols, and data sharing (Moran et al. 2014).

The network concluded that Americans' social outcomes and behavior are so situation- and place-specific that it is practically impossible to use widely dispersed national samples of populations to draw conclusions about processes in any one place. Although we can describe average educational attainment across 4th graders in the United States using data from the National Assessment of Educational Progress, the averages do not elucidate the social and cultural processes underlying educational underachievement in specific school districts. They can suggest general trends, but they cannot identify at-risk counties or school districts and the causes of underachievement. Populations tend to be spatially clustered by characteristics. This has implications for important issues of national concern. For example, the United States has always been characterized as the land of opportunity, where anyone can, through effort, succeed and attain the American dream. Yet a recent study showed that one out of four children raised in the middle class has slipped downward by their early 40s (Acs 2011); even more surprising is the finding that the chances for upward mobility and its maintenance depend specifically on where people live (Olinsky and Post 2013; Chetty et al. 2013). Similarly, a University of Washington study revealed that life expectancy for American males (and females) varies by up to 18 years, depending in which American county they reside (Institute for Health Metrics and Evaluation 2013). That is a remarkable range, and social scientists are just now mapping the social indicators to document and put it into context. Research focusing on context has the promise of pointing to pockets of concentrated disadvantage and poor health, where resources can be targeted to do the most good.

Why is there so much place-based variability in such outcomes as health and mobility? Social policy, economic conditions, race/ethnic/immigrant composition, and the size of the middle class in a city provide a context in which individuals can thrive or stagnate, leading to important questions about the roles of government, nonprofits, and for-profit organizations in fostering opportunity and mobility processes, and the impact of their policies on individuals, families, and neighborhoods. When and how does neighborhood context matter? What are the consequences of economic and social conditions and change in those conditions for individual economic opportunity and mobility? The social science platform of regional data centers that we envision can address such questions and others central to our understanding of who we are as a people and a nation.

We need a new national framework or platform that is both scalable and flexible for addressing these new challenges, one that allows for rapid response to local crises such as the devastating impact of tornadoes and hurricanes or social unrest, and those provoked by prolonged regional drought or economic decline. We need a platform to address national crises such as the declining middle class and why in many schools across the country our children are not learning optimally. Over the past few years, experts have called for increasing our capability in cyber-infrastructure for the social, behavioral, and economic (SBE) sciences. We propose building this national network of social observatories to ensure that the

SBE sciences have an effective scientific infrastructure to contribute to and collaborate with other sciences in addressing questions of national importance.

What Should a National Network of Regional Data Centers Look Like?

Such a network needs to be capable of representing the people and the places where people live both in the aggregate and in the fine detail—detail that captures local and regional differences. In doing so, the network would be able to paint a picture that is representative of the nation's population and a picture of population dynamics as detailed as is currently captured by our national surveys and research infrastructure. However, it would go far beyond the national surveys and would also be a place-based sample. Unlike most existing research platforms, this place-based capability will ensure that we understand not only the urban places where the majority of the population lives, but also the important medium and low density places that represent a vast majority of the communities and land area in the nation. To do so, the observatories would identify several hundred census tracts to be systematically studied over time and space by regional data centers spread across the country. Unlike the census, taken every 10 years, such regional data centers will be able to provide a continuous stream of information for the nation and thereby better address the dynamics of change in our society, allowing researchers to be able to quickly see how national policies affect local places. This is a totally different effort from that undertaken by our national surveys. It is not meant to replace them but rather will both offer a broader national picture of the population and also deepen our understanding of the population in places across the country using fine-grained methods. By addressing the context of individual activities and decisions, social observatories would provide a more complete understanding of socioeconomic success and failure in our society and what we might do to promote the former.

The observatories would study the social, behavioral, and economic experiences of the population and its physical and environmental context in fine detail. They will do so by using complementary methods, including ethnography, experiments, surveys, observations, geographical information systems, systems science, records searches, and historical and archival methods. These observatories will work closely with local and state governments to gain access to administrative data that will provide not a sample of the population of those hundreds of tracts, but complete records on the whole of the population in those tracts, thereby ensuring a depth of understanding, and integration of knowledge, heretofore never achieved in analyses. A network of regional data centers would do all this while at the same time being less invasive—and the data less prone to declining participation and response rates than national surveys—because they would be more closely tied to the local community through agreements with local private and public institutions and not need to rely on telephone calls. No matter how the nation changes or where its people move, the observatories would be able to describe how people and places change over time.

To date, the best neighborhood studies feature specific cases, such as Chicago or New York City, where substantial investments have been made to create systems of linked data such as we propose for these regional centers. They have been created because our larger cities have recognized that they need better data to serve their citizens while minimizing costs of such data acquisition and use. With a network of regional data centers, we can contemplate the possibility of a national sample of neighborhood contexts that can be studied at multiple levels and in multiple ways. Linked with health outcomes, it will be possible to consider the effects of, say, poverty while taking into account chemical exposures; it will also be possible to consider the effects of chemical exposures while controlling for poverty and other social characteristics of local contexts. There are virtually no studies to date that have done this.

Our Proposed Network

Our group, and earlier workshops on cyber-infrastructure for the social and behavioral sciences, has proposed the development of twenty to twenty-five regional data centers located across the United States. Each center would collect, organize, create, and disseminate data. These regional data centers or "social observatories" would follow about 400 census tracts over time and space from these twenty to twenty-five regions across the country. Working closely with local and state governments, they will access administrative data that will provide not a sample of the population of those several hundred census tracts, but complete records on all the population in those tracts (i.e., circa two million Americans). The centers would serve as data collection facilities wherein data are cleaned, linked, and made available for legitimate research purposes through a secure integrated data dissemination system. Although they would be charged with keeping data on their particular geographic region, some of these centers may have a national focus as well. These centers may conduct surveys, but they would also use data sources that until now have not been part of the toolbox of the social sciences, and they would connect these data to local context and place without losing the capacity to aggregate and serve as a national sample of people and places in the United States. Our vision is that, collectively, they would offer a nationally representative sample, one that was highly clustered so as to capture local context and variability.

A national framework provides an enormous advantage. First, it allows generalization across multiple contexts. The national framework permits comparison of variables and questions across multiple locations. It provides improved conceptual models that are not specific to place and can take into account variability. It ensures national representativeness. Second, the framework provides a rapid response capability. Over the observational period, at least some of the sites are likely to experience emergencies or crises. Because of the time dimension, the design will have an improved ability to disentangle causality. Third, having a decentralized structure permits each center to have a unique substantive focus.

Substantive foci of the regional data centers

Two major concerns about America's future are the adequacy of its physical infrastructure and the robustness of its economic structure. Many people worry that our transportation, environmental, water, sewer, educational, and even governmental infrastructure are outdated and in need of replacement. Yet efforts to move this agenda repeatedly fail to gain support in legislatures and among the public, despite the obvious benefits to business and citizens. People also worry that structural changes in the U.S. economy are making the nation less competitive in a global world and that the economy offers opportunities to some but not others. *The Economist* (2016) noted that the U.S. economy lacks vitality and competitiveness, and that it appears to be moving toward ever greater concentration of wealth, yet it is failing to benefit the larger population on which the economy depends. High profits are absorbed by ever more concentrated institutions, rather than being passed on to consumers or invested in innovation. This is a formula that reminds one of oligopolistic behavior: very high returns on capital, ever greater concentration, and control of prices in the hands of a handful of firms that cannot but lead to greater wealth concentration and inequality, according to the *Economist* article.

The major substantive foci of the data centers would initially be on questions of (a) change and adaptation and (b) opportunity and mobility, both broad questions that require data linkages and granularity in data sources. Because they would focus on place and context as well as people, researchers could identify the kinds of investments in infrastructure that provide the greatest opportunities for improvements in well-being with the fewest barriers. For example, one of the growing challenges for poor neighborhoods has been the exodus of grocery stores and therefore access to fresh food at reasonable prices. Creating opportunities for businesses to provide better access to healthy food can be investigated as a way to improve the lives of people who may be at an economic disadvantage. Regional data centers could also provide information on the organizational structures of communities that may facilitate or hinder appropriate adaptation to ongoing economic, social, and environmental change. This detailed understanding can inform public agents about how local economies might need to be reconfigured to better compete for jobs, for example. Of course, identifying needs in particular areas of the country creates the potential of having to ameliorate or reverse structural inequities and could lead to conflict among local groups. However, this does not mean that we should not move ahead to identify such needs and seek solutions.

What kinds of data would be collected?

To study local contexts such as communities and neighborhoods, we need spatially referenced administrative data, GPS-enabled cellphone data on the movements of individuals through their day, social media data, remotely sensed and observational data such as Google Street View, and survey data. These new data will be rich in detail. But while detail is important, the key to their use for social and behavioral research is linking them across different levels. Linking individual information to administrative data or to other characteristics of communities in which individuals live, linking information on social media activity to health or other events occurring within the area, and linking medical records with housing and health data are only some such examples.

Designed properly and operated efficiently, these networked regional data centers will provide a nimble platform to incorporate changing sources of information that are being created by social media companies and on other media platforms such as cellphones. The goal is to gain access to the new forms of communication used by the nation to understand how they transform how people think and what motivates them to act in certain ways, and how they construct virtual and real social networks and communities. The task here will be to improve the granularity of data; provide in-depth context to data; and address issues of social, time, and spatial scales. New cutting edge approaches such as data trawling and web scraping will produce detailed accounts of movement, social networks, and other forms of community building that require interpretation by bringing social theory and history to inform the analysis of tweets and other data moving across cyberspace.

Having these networked regional data centers will transform how the SBE sciences go about their work; they will encourage the integration of the SBE sciences, rather than promoting the fragmentation that we have experienced since the 1960s. The latter was a necessary phase to achieve greater depth through specialization but has over the years had the effect of making it ever more difficult for SBE scientists to share methods and approaches to address issues of national importance. The regional data center network will explicitly promote what is now a broad call from the National Academy of Sciences to integrate the social and physical sciences to address issues of importance with the best tools available without regard for disciplinary origins. Regional centers across the nation, with the explicit charge of ensuring that teams of scientists are working together around questions of national interest, will help to integrate the sciences and serve the nation better by providing diagnostic and policy-relevant solutions at a variety of scales from local to state to national to international issues. Although there are critical issues of privacy to be addressed in this geocoded world, the observatories will be a place where these concerns can be addressed systematically and lead to the creation of standards for ensuring privacy of sensitive information.

Regional and local data centers are already happening

A number of communities across the United States are developing collaborative regional data gathering efforts to document the linkages between people and place that go beyond specific city or state boundaries. The National Neighborhood Indicators Partnership (Kingsley and Pettit 2011) is active in more than thirty-seven cities. This partnership collects and shares data to better serve their communities and learn from one another. In addition to this existing network, a community of scholars has been working on individual elements of an ambitious network of regional data centers to ensure that the American people have available in a timely fashion nationally scaled and locally relevant information to make better business, health, education, and other important decisions. These include scholars at the New York Academy of Medicine and the New York City Department of Health and Mental Hygiene, the University of Pennsylvania, Chapin Hall–University of Chicago, the University of Colorado–Boulder, Portland State University, the University of Dallas, the Ohio State University, and American University, to name just a few examples. Cities such as Chicago have built impressive spatially explicit data bases that allow for quicker responses to social needs, and have provided a public portal so that citizens can engage the government to be more responsive and can be engaged with what happens in their city. We have seen a proliferation of these efforts across the nation. There are efforts to articulate some of these endeavors, but a larger and more systematic effort is needed to ensure that these efforts coalesce and provide a more complete picture of both local and national processes.

Methodological advances: Linkages across data

The important challenge to data collection for the foreseeable future is linking diverse data in a way that is cumulative, accurate, and accessible to a broad range of researchers. For SBE research, it is critical to develop new methods that can link persons and information about them and their environments across different data platforms. The proposed regional data centers would undertake the challenging task of linking a broad array of information—from administrative data (local and state and regional), to social media (Twitter, Facebook), to census and other surveys, to ethnographic data, to data from experiments such as randomized controlled trials—to address how different human communities make decisions. This is the issue that was addressed by the conference convened by the SOCN in 2016 on New Data Linkages, and that this volume addresses.

How This Volume Will Move Us Forward

The NSF SOCN sponsored a conference in the Washington, DC, area on March 24–25, 2016. The purpose of this conference was to bring together researchers involved in different regional data collection and linking efforts to (1) promote synergies across projects and (2) explore what types of issues have arisen that could be facilitated by a regional system of data centers. The call for papers was issued in spring 2015 across a number of academic disciplines, including Demography, Sociology, Economics, Psychology, Anthropology, Geography, Hazards and Environment Risk, Political Science, and Statistics. From this call, we gathered eight different research teams to discuss their work and explore potential collaborations across these projects. Besides the principal investigators of these projects, we invited members of the SOCN and guests from funding agencies such as the National Institutes of Health, NSF, the National Academy of Science, and private foundations to serve as discussants and observers.

This volume has the following structure: This article, the introduction, documents our thinking about how to develop a network of regional centers into a national platform capable of serving the SBE sciences and how several research groups are already making progress in linking data. In the following three sections, authors give concrete examples of how linked data are advancing research in (1) community characteristics and quality of life, (2) individual and community factors and health, and (3) change and adaptation and disaster planning. Although these topics are not the only issues in which such a network could be actively involved, community quality of life, health and health care, and adaptation to immigration and climate change represent important areas of concern for families and for future public policy discussion in the United States, and all are issues that have attracted a great deal of attention and on which solutions remain incomplete and so far unsatisfactory. We then offer our conclusions and suggestions for future research.

Community characteristics and quality of life

In the first article of this section, Michael Bader and colleagues report on a new initiative in the Washington, DC, area that brings physical characteristics of communities into research on behavior. His team addresses how to link data from Google Street View to better assess physical infrastructures that can facilitate or hinder the mobility of the aged. Second, Christopher Browning and colleagues, a team from Ohio, summarize their research on adolescent behavior through the use of new tools such as smartphones that map activity spaces instead of only neighborhoods, and link individual and activity space data with administrative data. Their article addresses the characteristics that affect exposure to violent locations, which could threaten mental/physical health. Finally, Amy O'Hara, Rachel Shattuck, and Robert Goerge report on both U.S. Census Bureau and Chapin Hall–University of Chicago initiatives to integrate data sources for improved research on families. They link federal surveys with federal and state administrative data to better measure families and households, obtain more extensive information on families, evaluate survey coverage and accuracy, and evaluate participation in social welfare programs.

Individual and community factors and health

The first contribution in this section—from Hongying Dai, Brian Lee, and Jianqiang Hao, a team from the Children's Hospital and the University of Missouri—uses novel linkages of Twitter data, health survey data, and socioeconomic data from the U.S. Census Bureau to predict community asthma burden. Next, a team from the Population Council extends the study of health to the use of modern contraception in African communities. Jean Digitale and colleagues describe how individual and community factors are associated with contraceptive use among young women in Malawi and their selection of a contraceptive provider. The authors link data from a survey of youth with data from a survey of family-planning service providers. The third contribution is from a team at the University of Dallas and University of Texas Southwestern Medical Center, which examines the health of residents of low-income areas in the city of Dallas. Tammy Leonard, Amy Hughes, and Sandi Pruitt examine the coping strategies of families experiencing a health shock, by linking medical record data with community service administrative data and housing appraisal data.

Adaptation and disaster planning

Coming from the Population Studies and Training Center at Brown University where she has been studying the consequences for New Orleans of Hurricane Katrina, Elizabeth Fussell and her colleagues offer a study of how past population trends, population density, cumulative weather-related losses, and weather events intersect at the county level to influence future population change. Also in this section, Elyakim Kislev studies what will likely be one of the major challenges for the United States in the not too distant future: the integration of European immigrants whose country of origin is Africa or the Middle East. Kislev demonstrates how a variety of linked data can be used to track the mobility of immigrants in Western Europe, examines the characteristics of those who subsequently immigrated to the United States, and compares the successes of these immigrants with different origins to U.S. natives.

In the concluding article to the volume, Barbara Entwisle, Sandra Hofferth, and Emilio Moran offer thoughtful reflections on the articles in this volume and how and in what ways we might move forward to achieve the promise presented by data linkages within the context of a national network of regional centers that can enhance the SBE sciences and, in doing so, serve society.

We invite readers to join us in working toward the advancement of interdisciplinary science by taking on the challenge of linking relevant data and utilizing innovative methods to elucidate social dynamics and solve challenging problems all around us now and in the foreseeable future.

Note

1. See materials at www.socialobservatories.org.

References

Acs, Gregory. 2011. *Downward mobility from the middle class: Waking up from the American dream.* Washington, DC: Pew Charitable Trusts.

Chetty, Raj, Nathaniel Hendren, Patrick Kline, and Emmanuel Saez. 2013. *The economic impacts of tax expenditures: Evidence from spatial variation across the U.S.* Cambridge, MA and Berkeley, CA: Equality of Opportunity Project, Harvard University and the University of California, Berkeley. Available from http://www.equality-of-opportunity.org/.

Economist. 26 March/1 April 2016. Too much of a good thing: Why high profits are a problem for America, 23–28.

Foster, Ian, Rayid Ghani, Ron S. Jarmin, Frauke Kreuter, and Julia Lane. 2017. *Big data and social science: A practical guide to methods and tools.* Boca Raton, FL: CRC Press

Institute for Health Metrics and Evaluation (IHME). 2013. *The state of U.S. health: Innovations, insights, and recommendations from the Global Burden of Disease Study.* Seattle, WA: IHME.

Kingsley, G. Thomas, and Kathryn L. S. Pettit. 2011. Quality of life at a finer grain: The National Neighborhood Indicators Partnership. In *Community Quality-of-Life Indicators: Best Cases V*, eds. M. J. Sirgy, R. Phillips, and D. Rahtz, 67–96. New York, NY: Springer

Moran, Emilio F., Sandra L. Hofferth, Catherine C. Eckel, Darrick Hamilton, Barbara Entwisle, J. Lawrence Aber, Henry E. Brady, Dalton Conley, Susan L. Cutter, and Klaus Hubacek. 2014. Building a 21st century infrastructure for the social sciences. *Proceedings of the National Academy of Sciences* 111 (45): 15855–56.

Olinsky, Ben, and Sasha Post. 2013. *Middle out mobility: Regions with larger middle classes have more economic mobility.* Washington, DC: Center for American Progress.

The Promise, Practicalities, and Perils of Virtually Auditing Neighborhoods Using Google Street View

By
MICHAEL D. M. BADER,
STEPHEN J. MOONEY,
BLAKE BENNETT,
and
ANDREW G. RUNDLE

In-person audits to collect data on neighborhood characteristics offer opportunities to study the mechanisms that link neighborhood conditions to unequal outcomes for individuals and communities, but the expense and logistical difficulties associated with conducting neighborhood audits have limited their use. The images collected by Google Street View provide a promising alternative for researchers to measure neighborhood environments across cities and to examine how neighborhood conditions vary across a wider geographic scope. We describe the benefits of using "virtual" neighborhood audits and discuss the practicalities of collecting data from virtual audits. We provide an example of individual- and neighborhood-level inequality in the distribution of disorder for older adults across four cities: New York, San Jose, Philadelphia, and Detroit. Despite the promise of virtual audits, they also introduce perils that must be addressed as research progresses; we introduce and discuss those perils here.

Keywords: Google Street View; systematic social observation; neighborhood audit; aging in place; neighborhood effects

Evidence continues to mount that demonstrates the relevance of neighborhood conditions to the health and well-being of residents

Michael D. M. Bader is an assistant professor of sociology at American University in Washington, DC. He studies how patterns of neighborhood change have evolved since the civil rights movement, processes that perpetuate spatial inequality, and measurement of neighborhood environments.

Stephen J. Mooney is a postdoctoral fellow at the Harborview Injury Prevention & Research Center at the University of Washington in Seattle. His research focuses on spatial and contextual determinants of health behaviors, particularly physical activity and injury.

Blake Bennett has 10 years of research experience primarily in the field of toxicology. His research combines methodologies from the fields of environmental science and public health to assess how differences in urban form influence health.

Correspondence: bader@american.edu

DOI: 10.1177/0002716216681488

in those neighborhoods (e.g., Diez Roux and Mair 2010; Diez Roux 2003; Sampson 2012). As study designs have come closer to establishing a causal relationship between neighborhoods and economic and health outcomes of residents, greater attention should be devoted to the mechanisms through which neighborhoods create unequal outcomes (Chetty, Hendren, and Katz 2016; Ludwig et al. 2011, 2012; Sharkey and Sampson 2010). The unequal distribution of physical conditions across neighborhoods presents one potential mechanism that can help to explain racial and economic inequality.

Over the past three decades, social scientists and public health researchers have developed methods to measure the physical conditions of neighborhoods (Pikora et al. 2002; Reiss 1971; Sampson and Raudenbush 1999; Taylor, Gottfredson, and Brower 1984). Researchers train raters to collect data on specific items related to the physical conditions visible from streets (e.g., condition of sidewalks, presence of graffiti) and then send those raters to prespecified locations within neighborhoods. These neighborhood audits or systematic social observations generate data that can then be aggregated to create neighborhood-level measures (Mujahid et al. 2007; Raudenbush and Sampson 1999). Because in-person neighborhood audits are expensive, most studies incorporating this method have been limited to one study site. This narrow geographic scope has limited the generalizability of findings because physical features of neighborhoods vary across regions, meaning that the *associations* between physical features and outcomes might also vary.

Google's Street View product offers an opportunity to cost-effectively measure neighborhood conditions across metropolitan areas.[1] Street View displays photographs with detailed geographic data from neighborhoods across the world. One can conduct audits analogous to in-person neighborhood audits using the photographs compiled in Street View. The systematic collection and analysis of Street View data have been used to assess the relationships between neighborhood conditions and individual outcomes (Odgers et al. 2012) and to study the relationships between social and economic change and changing physical conditions of neighborhoods (Hwang and Sampson 2014).

While Street View and other geographic "big data" offer new opportunities (Mooney, Westreich, and El-Sayed 2015), these data also create new problems. We illustrate how we used Google's Street View to describe the variation in the associations between neighborhood physical disorder and both racial and economic composition for aging residents in four cities from different regions in the United States: New York, San Jose, Detroit, and Philadelphia. We analyzed these data as a simple example to show how researchers and policy-makers can make use of geographic "big data" to understand an important contemporary problem in society: the physical infrastructure in neighborhoods that house America's aging population. We conclude by offering future steps that should be taken when using Street View for research.

Andrew G. Rundle is an associate professor of epidemiology at Columbia University's Mailman School of Public Health. His research focuses on the determinants of sedentary lifestyles and obesity and the health-related consequences of these conditions.

The Benefits and Costs of Neighborhood Audits

Neighborhood audits (also called systematic social observations) have a long history in social science and public health research (Perkins and Taylor 1996; Pikora et al. 2002; Reiss 1971; Sampson and Raudenbush 1999). Researchers have used audits to consider how the physical and social environments in which we live affect behavior and outcomes including neighborliness, crime, obesity, and physical activity (Browning et al. 2010; Perkins et al. 1990; Sampson and Raudenbush 1999; Taylor et al. 1984; Zhu and Lee 2008). Observing neighborhood characteristics can help researchers to isolate the influence of urban design and physical conditions net of respondents' own reports of their environment (Kirtland et al. 2003). In doing so, neighborhood audits help researchers to avoid the methodological problem of "same source bias" that arises when measurement error is correlated between dependent and independent variables (Duncan and Raudenbush 1999). In addition, directly measuring neighborhood features helps researchers to detect social biases, like racial prejudice, that affect respondent reports about neighborhood conditions (Sampson and Raudenbush 2004).

To create neighborhood-level measures, researchers rely on statistical methods to aggregate the data. These methods borrow from psychometric measurement models that combine items into scales measuring latent traits (Mooney et al. 2014; Mujahid et al. 2007; Raudenbush and Sampson 1999).[2] In neighborhood models, these latent traits comprise concepts such as "disorder" or "walkability," which no single item can effectively measure. Therefore, "ecometric" scales estimate these latent traits by aggregating multiple items (e.g., litter, graffiti, boarded-up buildings) measured at multiple sampled locations within a sample of neighborhoods (Raudenbush and Sampson 1999). Researchers have, in recent years, investigated ways to use the spatial autocorrelation between neighborhoods to improve neighborhood measures (Bader and Ailshire 2014; Savitz and Raudenbush 2009).

These methods can be effective only with both a sufficiently sized sample of locations within neighborhoods and a sufficient sample of neighborhoods. Unfortunately, doing audits correctly and collecting sufficient samples within and between neighborhoods also increase the cost of studies. A large and representative sample of observations in neighborhoods must be added to a large and representative sample of people to measure the outcomes of interest.

Conducting audits also increases the logistical difficulties of running studies. Matching audits to the correct geographic location and ensuring the safety of auditors both factor heavily into logistical constraints (Carter, Dougherty, and Grigorian 1995). Having outsiders enter neighborhoods with clipboards or other devices could also breed suspicion or resentment, which could undermine community trust and impede research. The logistical difficulties of conducting audits substantially increase when conducting studies across multiple sites since local staff must be trained, logistical services must be duplicated, and community trust must be fostered.

Sampling and logistical difficulties generally limit the scale of research that uses neighborhood audits. While neighborhood audits provide the benefits of statistically independent neighborhood measures that can help to uncover the mechanisms through which neighborhoods affect individuals and society, the costs often make their use impossible. The costs increase the most when trying to deploy neighborhood audits across multiple sites even though interregional studies offer the most potential for scientific insight into neighborhood conditions and their related outcomes.

The Promise of Research Using Google Street View

Cost-effectively auditing neighborhood environments using Street View

The large and growing worldwide coverage of streets included in Google Street View's database can reduce the financial and logistical problems associated with collecting neighborhood audits. To create its Street View product, Google uses cars carrying specially equipped cameras to travel streets and photograph the streetscape (Anguelov et al. 2010). More recently, bicycles and other modes of transportation have also been used to capture pedestrian- and bike-accessible areas. The Google camera captures the precise geographic location of each image. Google then stitches together the photographs for display on the Google Maps website. The website allows users to take a virtual "walk" down the street and view an uninterrupted scene of photographs.

The cost of traveling to neighborhoods makes up the largest expense of neighborhood audits and typically limits research to a single site. For example, in a pilot test that we conducted in summer 2009, we found that conducting "virtual audits" using Street View reduced the amount of time required to conduct audits using the complete Pedestrian Environment Scan Data (Clifton, Smith, and Rodriguez 2007) on seventy-four street segments in New York City by 70 percent. All of these time savings resulted from eliminating travel to street segments, since collecting observational data on physical conditions of street segments in person took approximately as long as virtually walking down the street in Street View.

These savings do not come at the cost of data quality. The data obtained using Google Street View provide valid and reliable measures of many items. A host of studies has established validity across different settings by showing high levels of "intersource" agreement between virtual audits using Google Street View and in-person audits including in New Orleans (Curtis, Duval-Diop, and Novak 2010), New York City (Rundle et al. 2011), Chicago (Clarke et al. 2010; Hwang and Sampson 2014), and Atlanta (Vargo, Stone, and Glanz 2012). A sample of 120 street segments in the United Kingdom showed the potential for moderate to high levels of interrater reliability using Google Street View (Odgers et al. 2012), while a study of 150 street segments found similar levels of interrater reliability in the United States (Bader et al. 2015). Since these two studies used similar items, using Google Street View could open possibilities for international comparisons in addition to national ones.

The specific types of items auditors can reliably rate using Street View imagery include features of roads and buildings, and urban design infrastructure. Because the camera sits on a car driving down the road, items designed to be seen by drivers rank among the most reliably measured items (Bader et al. 2015). Raters also reliably rate features of the built environment, including how land is used and the physical condition of buildings because the objects are large, and thus visible, and change infrequently (Bader et al. 2015). Items with high levels of temporal variation or small tend to be less reliably rated. Sounds, smells, and the feel of being on the street cannot be rated at all.

By reducing the cost of measurement, virtual audits using Street View also create opportunities to study how neighborhoods change over time. In 2014, Google introduced a feature to Google Maps that allows you to look at past imagery; this allows observers to examine how neighborhood features change through construction or after natural disasters (Shet 2014). Comparing quantitative observations over time also allows researchers to investigate mundane, but potentially influential, change, such as gentrification. For example, Hwang and Sampson (2014) compared virtual audits using Street View images taken from 2007 to 2009 to in-person observations conducted in 2002 and 1995 to assess the neighborhood characteristics associated with gentrification.[3]

Leveraging the Google Maps application programming interface to conduct neighborhood audits

With the Google Maps application programming interface (API), which includes methods for retrieving Street View data, you can create applications that allow cross-site studies. Large representative samples of neighborhoods can be drawn with little expense. The API allows researchers to create samples using a single protocol that can also be used to gather precise geographic data along with the data auditors collect. With a global library of images, researchers can study entire countries or even make cross-country comparisons using Street View. Researchers, including our team, have used the API to develop online applications that help researchers to use Google Street View for neighborhood audits without requiring a substantial amount of computer programming expertise. These tools, including the "Systematic social observation inventory – Tally of observations in urban regions" (SSO i-Tour; adaptlab 2012) and the Computer Assisted Neighborhood Visual Assessment System (CANVAS; Bader et al. 2015), provide researchers with the opportunity to collect data on large samples of neighborhoods. These applications can also provide tools to help assess the progress and reliability of ratings as studies are conducted.

Practicalities of Virtual Neighborhood Audits: Assessing Disorder among Aging Residents

We demonstrate how virtual audits can be combined with census data to assess policy related to the aging population in the United States. The leading edge of the baby boom cohort born after World War II has reached retirement age.

Medical advances that prolong life in old age have led to a growing number of aging Americans. As they age, almost nine in ten Americans 65 and older express a desire to continue to reside in their own home and community (Kennan 2010). Advocates and policy-makers have attempted to address this desire with policies and programs to help older adults in the United States to "age in place" (Bayer and Harper 2000; Farber et al. 2011). States have developed legislation to help residents age in place in part because aging in place can save $1,500 per year in Medicare and Medicaid spending relative to nursing home care (Marek et al. 2012). But the ability to stay in one's home requires more than the will to do so. Aging in place requires that the elderly's neighborhoods provide environments conducive to active and engaged living (Clarke and Nieuwenhuijsen 2009; Clarke et al. 2008). Older adults need to feel comfortable in their environment and require the infrastructural support necessary to do so.

Policy recommendations have minimally focused on neighborhood environments. A joint report by the National Conference of State Legislatures and the AARP Public Policy Institute, for example, focused on three areas for legislation: land use planning, public transportation access, and affordable housing with appropriate modifications to accommodate aging bodies (Farber et al. 2011). While important, these policy recommendations overlook other aspects of neighborhoods that affect older Americans. Physical neighborhood conditions need to be just as accommodating for aging bodies as do home environments. Older adults might be particularly influenced by neighborhood physical disorder that would affect their engagement with friends and neighbors outside of their homes (Klinenberg 2003). The physical conditions of neighborhoods surrounding aging adults provide an example of the advantages associated with using long-term observations of neighborhoods and populations.

We combined data collected using virtual neighborhood audits in four cities with the 2010–2014 American Community Survey five-year estimates at the tract level to assess racial and economic inequalities in exposure to neighborhood disorder among elders. Our assessment examined two different questions: What are the different levels of exposure among groups of elders (i.e., between poor and nonpoor elders and among black, white, and Latinos elders)? and What neighborhood characteristics correlate with neighborhood disorderliness?

Project team

Conducting virtual audits requires people to fulfill several different roles. Every project needs a leader or leadership team that develops the research questions that can be answered using the data observed from the neighborhood audit. Research teams require a trainer and manager who oversee the raters who actually use the Street View images to collect observational data at specific street intersections or segments. Someone on the team also needs to fulfill the role of a spatial analyst trained in geographic information systems to understand and appropriately analyze the geographic characteristics of the data.

The background knowledge of raters and geographic context of the study should be considered when developing training materials. Auditors often need more help on questions that relate to urban design. For example, auditors that we trained before conducting a national reliability sample were often confused by differences between traffic calming devices such as chicanes, traffic circles, and curb extensions. Building styles vary greatly nationally and internationally, so training materials should be developed based on the geographic regions covered in the study. For example, a study using virtual audits in a single metropolitan area should tailor the training materials to that metropolitan area, whereas a study assessing national locations should use training materials with examples from as wide a cross-section as possible. Modifying items for use across geographic locations might be necessary to reliably assess characteristics in large study areas.

Work flow

Our data collection and analysis followed four steps. In the first step, we *sampled locations* on a regular geographic grid across each of the four cities. We then loaded the Street Viewable segments into CANVAS, a web application that we developed to allow raters to rate streets using the Google Maps API and web-based forms (Bader et al. 2015). We then used the Google Maps API to identify whether a "Street Viewable" street segment (i.e., a street segment that included Street View imagery) existed near that point. In the second step, raters *virtually audited* the street segments for a number of measures, including items measuring physical disorder. Third, we downloaded data from the CANVAS application and *created a physical disorder scale* that aggregated the neighborhood disorder measures. Finally, we included the neighborhood scales created from the virtual audit data to *analyze disparities by race and class*.

Sample locations. We used a regular one-kilometer grid starting at a random coordinate in the city to sample locations in Street View. These data were initially collected to assess child outcomes from a different study, so a secondary half-kilometer grid was overlaid to increase measurements in areas with a high density of study respondents. We then used the sampling function built into the CANVAS application to find a Street Viewable segment at the sampled point in the Google Maps API. If the CANVAS application could not find a Street Viewable segment at the location, it would look in up to five random locations near the sampled point to find a segment (for details, see Mooney et al. 2014). Table 1 reports the number of sampled and Street Viewable segments in each city.[4]

Virtually auditing street segments. Students, most of whom were undergraduates, were trained to use the CANVAS auditing system to rate the physical characteristics of intersections and street segments. Data collection took place from June 2012 to June 2013 and mostly occurred in an office where several dual-monitor workstations were set up for their use. When the auditors logged into CANVAS, the application assigned auditors to audit intersections and street segments across the four cities. The CANVAS application placed starting and ending

TABLE 1
Number and Proportion of Street Viewable Segments in Four Cities

City	Sampled	Street Viewable	Proportion Street Viewable
New York City	541	532	0.98
San Jose	324	289	0.89
Philadelphia	533	503	0.94
Detroit	517	502	0.97
Overall	**1,915**	**1,826**	**0.95**

markers on the Street View imagery as way-finders for the rater to follow, and raters were assigned to collect data (e.g., presence of graffiti on buildings) by observing the right side of the segment as they virtually "walked" from the starting marker to the ending one. Raters also were allowed to indicate whether their view was obstructed by parked cars, foliage, or other impediments and also whether the images permitted rating at all. When a rater finished collecting data on a street segment, CANVAS loaded the Street View imagery for the next assigned location from the sampling plan into the rater's browser window.

Creating a neighborhood physical disorder scale. After raters completed auditing all the viewable street segments in the sampling frame, we downloaded the data from CANVAS and constructed a physical disorder scale using item response theory analyses following the strategy of Raudenbush and Sampson (1999). This physical disorder scale comprised data for nine items: the presence of litter, bottles, graffiti (including that painted over); abandoned cars; buildings in poor repair; burned-out buildings; abandoned buildings; bars on windows; and vacant land. We combined the fitted value of each segment's scale score along with the segment's random error to measure overall physical disorder at each sampled street segment, in each of the four cities (Mooney et al. 2014).

We then used a geostatistical method known as "kriging" to interpolate the value of the disorder scale score at all street segments throughout the city based on the spatial correlation between the disorder scale score at the sampled locations. For each of the four cities, the starting and ending coordinates for every street segment were projected onto the appropriate state plane, and then kriging was used to estimate physical disorder scale score values on an approximately 300-by-300 foot grid in each city.[5] A final census tract–level physical disorder measure was then created by identifying the 300-by-300 foot grid points that fell within the 2010 U.S. Census tract boundaries and by taking the mean of the physical disorder scale scores estimated at those points.

Kriging provides several benefits to researchers studying neighborhood characteristics (Bader and Ailshire 2014). First, kriging provides an estimate and confidence interval at every location. Second, kriging provides an unbiased estimate of the interpolated value even if measured locations are irregularly sampled. For this reason, the oversample in particular neighborhoods will not bias the measures of disorder, though it will reduce the standard errors of estimates in areas that were oversampled. Finally, because the measure can be interpolated at any geographic point, researchers can use the resulting interpolations to measure neighborhood characteristics using different neighborhood definitions or scales. This is an advantage over measurement approaches that require a sample of neighborhood polygons to be drawn in advance of a study (Raudenbush and Sampson 1999; Savitz and Raudenbush 2009), which limits the degree to which the same data can be used across studies.

Analyzing unequal exposure to physical disorder. We standardized the neighborhood disorder scale based on the mean and standard deviation of the disorder values aggregated within tracts. Standardizing within cities means that comparisons are based entirely on the differences between groups within the same city instead of comparing the same value across cities. This prevents the relative size and distribution of groups across cities from influencing the findings and also allows baseline levels of disorder to vary across cities. Future studies with larger samples of cities could use this method to directly model both the variation within and the variation between cities.

To assess individual economic and racial disparities, we obtained data about the composition of the census tract from the American Community Survey five-year estimates from the years 2010 to 2014. For all analyses, we defined elders as those aged 65 and older. We identified whites as those who identified themselves as white alone and not Hispanic or Latino. We identified blacks as those who identified themselves as black alone and Latinos as people of any race who identified as Hispanic or Latino. To assess how neighborhood characteristics differ for older adults by race, we examined the proportion of black, white, and Latino residents aged 65 and older who lived in each neighborhood that fell into each quartile of neighborhood disorder within each city. The same procedure was used to assess the distribution of disorder among elders living in poverty.

To assess the correlations between physical disorder and neighborhood conditions, we regressed the level of physical disorder on the proportion of residents aged 65 or older who were white, black, and Latino, and by poverty rate. We weighted the regressions by the size of the elderly population living in each census tract, making the regression estimates representative of the average association between physical disorder and neighborhood characteristics for residents 65 and older in each city. We included several other variables in the regression that were potentially associated with physical disorder. We included the percentage of housing units that were vacant because high vacancy rates might reduce neighborhood vigilance and because vacant properties might become targets of disorderly behavior. We included the percentage of foreign-born residents to account for potential racial differences that might be attributable to immigration status. We included the percent of housing units occupied by owners because home owners might be more sensitive to disorder than renters. Finally, we included population density to assess land use and potential residential traffic in the neighborhood.

We do not know the true value of the disorder scale, only the interpolated value based on our kriging model. To account for this uncertainty, we created a series of ten simulations of disorder measures conditioned on the spatial covariance structure and values at the sampled locations. These conditional realizations are akin to multiple imputations, provide a sense of possible values across the city, and are particularly important to use since we estimate some areas in each city with more precision than others due to our two-stage sampling strategy. We therefore estimate a series of ten models and combine the parameter estimates and standard errors using Rubin's (1987/2004) corrections for imputed data. All analyses were conducted in R version 3.2.2, and multiple imputation corrections were conducted using the mix package (Schafer 2015).

Results

Unequal exposure to disorderly neighborhoods

Figure 1 displays cumulative bar plots of demographic differences in the exposure that elder New Yorkers have to physical disorder. The shading represents the percentage of the group living in neighborhoods that fall within each quarter of the overall distribution of disorder in New York City. The top bar shows the percentage of all adults aged 65 and older. Older adults were approximately equally distributed in each of the four quartiles of disorder, where a slightly higher proportion lived in the least disorderly neighborhoods and slightly lower proportion in the most disorderly neighborhoods.

Large racial and economic disparities existed, however, in the level of disorder that elders experienced. Two in five white adults 65 and older lived in the least disorderly neighborhoods in the city. Only 17 percent and 15 percent of black and Latino elders, respectively, lived in the least disordered neighborhoods while 25 percent and 39 percent of blacks and Latinos, respectively, lived in the most disordered neighborhoods. By comparison, only 9 percent of elderly whites lived in the most disordered neighborhoods. Poverty status also mattered: only 17 percent of adults in poverty lived in the least disordered neighborhoods, and 31 percent lived in the most disordered neighborhoods.

Demographic differences of San Jose residents ages 65 and older are plotted in Figure 2. Like New York, approximately equal proportions of elderly San Jose residents lived in neighborhoods falling in each of the four quartiles of physical disorder. Also like New York, large racial disparities existed as older white residents tended to live in far less disorderly neighborhoods than older black and Latino residents. Only 31 percent of white elders lived in neighborhoods falling in the most disorderly half of San Jose neighborhoods compared with the approximately 60 percent of both black and Latino elders. Unlike New York, however, older adults who found themselves in poverty were not particularly unequally distributed across neighborhoods with varying levels of disorder.

Figure 3 presents the data for Philadelphia residents aged 65 and older. While more older Philadelphia residents lived in Philadelphia's neighborhoods with low

FIGURE 1
Difference in Exposure to Quartiles of Physical Disorder, by Race and Class
in New York City

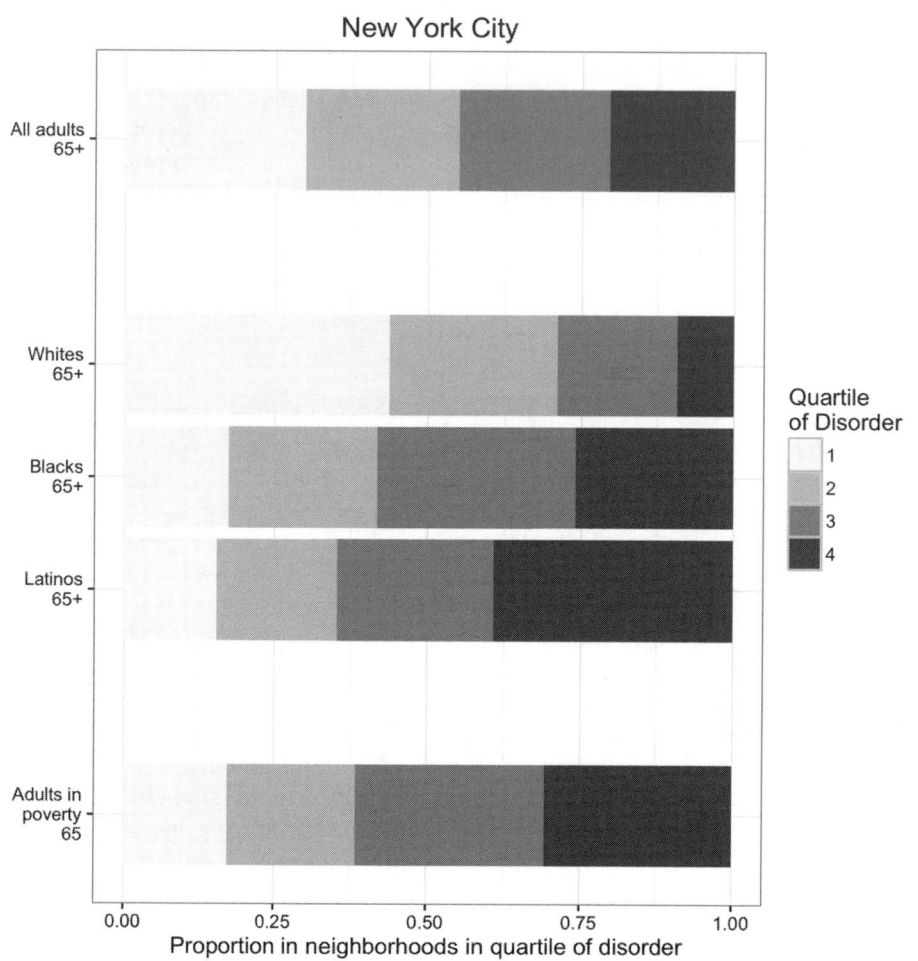

levels of disorder, race factored heavily into the level of disorder to which older Philadelphians were exposed. Only 6 percent of whites age 65 and older lived in the most disorderly quartile of Philadelphia neighborhoods, and only 18 percent lived in the next most disorderly quartile of neighborhoods. Together, about a quarter of whites lived in the most disorderly half of neighborhoods; this was smaller than the percentage of black and Latino residents who lived in the most disorderly *quarter* of Philadelphia neighborhoods, which was 37 and 38 percent, respectively. Three-fifths of blacks and two-thirds of Latinos ages 65 and older lived in the most disorderly half of Philadelphia neighborhoods, and only 12 percent of older black Philadelphians lived in the least disorderly quarter of Philadelphia's neighborhoods. Older adults living in poverty were slightly more

FIGURE 2
Difference in Exposure to Quartiles of Physical Disorder, by Race and Class in San Jose

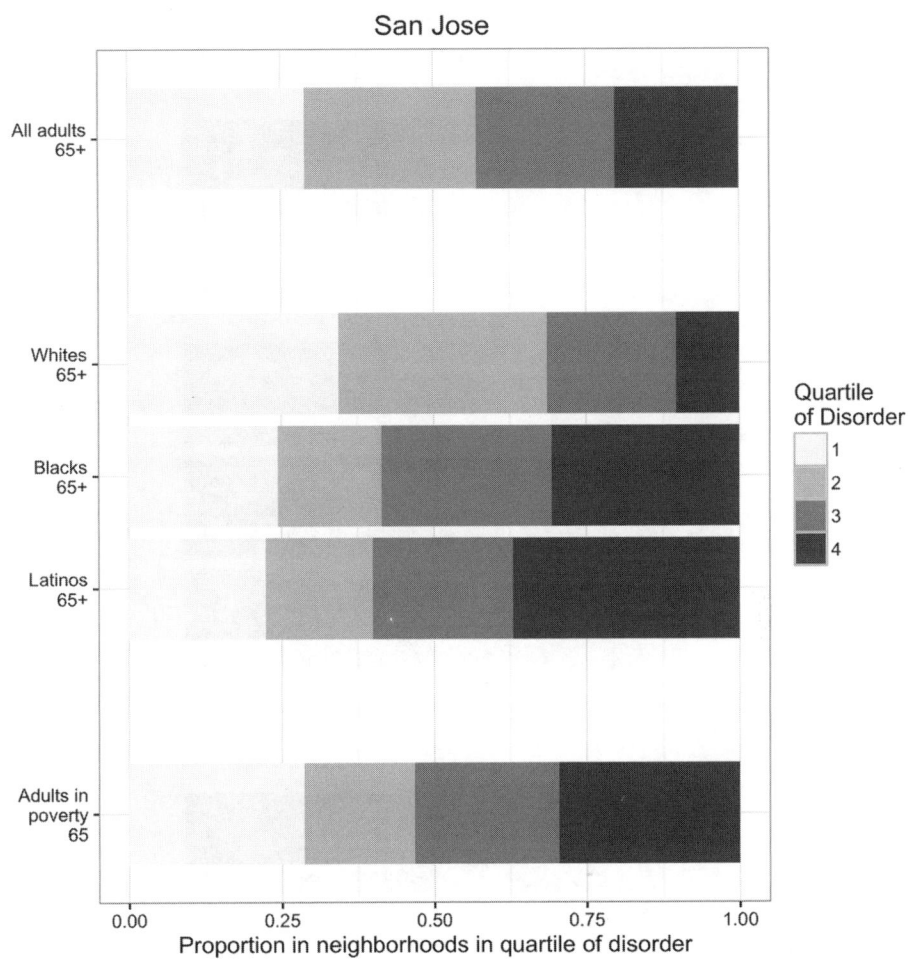

likely than older adults overall to live in more disorderly neighborhoods, but elders living in poverty were relatively evenly distributed across the range of disorder.

We plotted the results for Detroit in Figure 4. Unlike the other three cities, elder Detroiters tended to live in Detroit's least disorderly neighborhoods. Just more than a third of Detroit residents 65 and older lived in the least disorderly quartile of Detroit neighborhoods, and only 20 percent lived in the most disorderly neighborhoods. In addition, racial disparities among the elderly in Detroit were less pronounced. A larger percentage of black elders, 36 percent, lived in the least disorderly neighborhoods than of white elders, 28 percent. That said, a larger percentage of black elders, 22 percent, were exposed to the most

FIGURE 3
Difference in Exposure to Quartiles of Physical Disorder, by Race and Class in Philadelphia

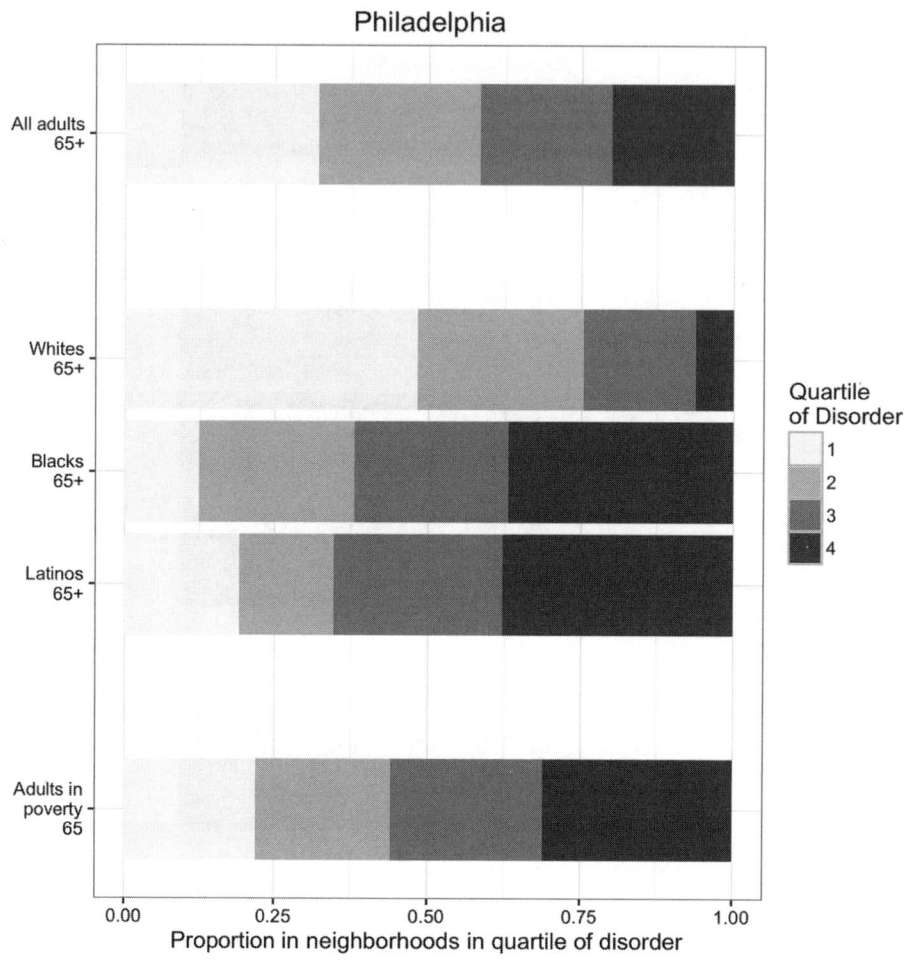

disorderly neighborhoods with only 13 percent for white elders. Exposure to disorder was higher among older Latinos in Detroit compared with older blacks and whites. Even so the racial disparities were not nearly as large as they were in the other three cities, as older Detroiters living in poverty were very evenly distributed among neighborhoods with varying levels of disorder.

In summary, residents ages 65 and older in all four cities tended to be approximately equally distributed across neighborhoods with varying levels of disorder. But racial disparities in exposure to disorder were very high between whites and blacks in three cities and whites and Latinos in all four cities. When we consider the potential for physical disorder to influence outcomes associated with aging,

FIGURE 4
Difference in Exposure to Quartiles of Physical Disorder, by Race and Class in Detroit

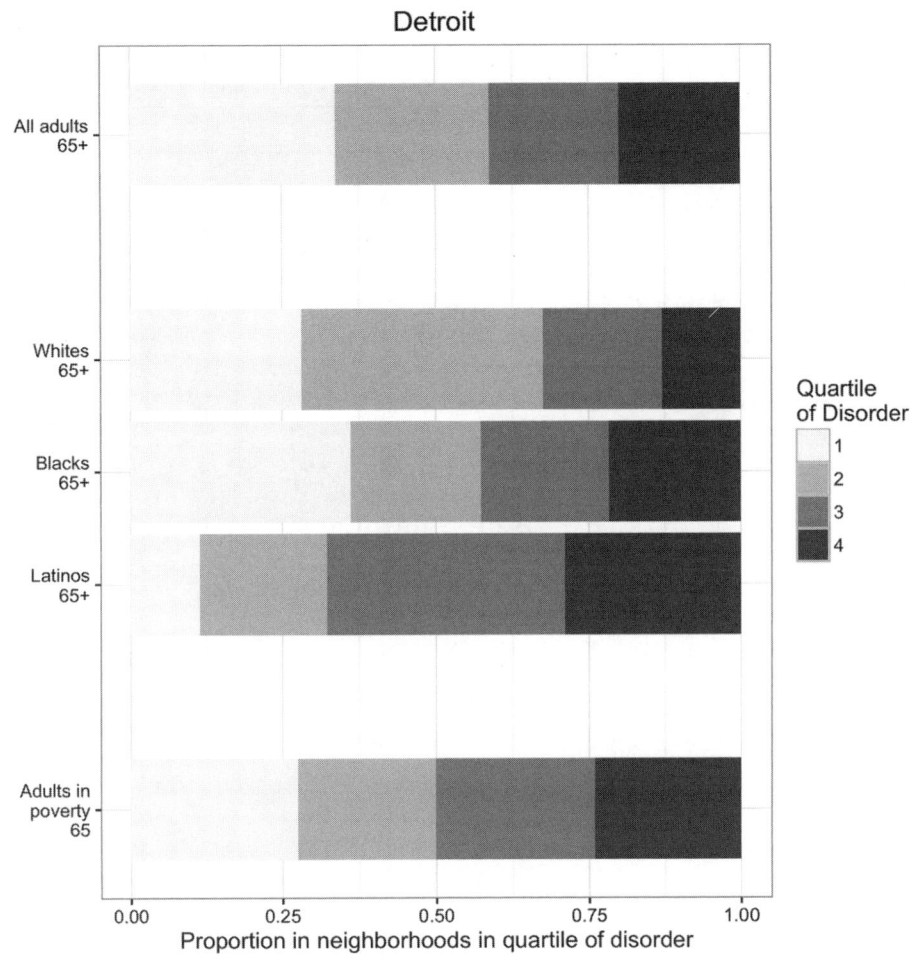

the effects will likely be concentrated among nonwhites. The higher level of disorder to which black and Latino elders are exposed could help to explain racial disparities in aging outcomes.

Neighborhood correlates of physical disorder

Table 2 reports the results of our analysis of the relationships between the level of physical disorder and the social and economic conditions in neighborhoods. We analyzed the results from each city independently with the dependent variable being the level of disorder standardized to the distribution of all tracts within each city. Coefficients represent the number of standard deviations of

TABLE 2
Parameter Estimates and Standard Errors of Ordinary Least Square Regressions of Neighborhood Physical Disorder on Racial and Poverty Composition with Covariates

	Race only		Poverty only		All covariates	
	b	SE	b	SE	b	SE
New York						
Pct. non-Latino black	0.01	0.00			0.00	0.00
Pct. Latino	0.02	0.00			0.01	0.00
Pct. in poverty			0.04	0.00	0.01	0.00
Pct. college educated (ages 25+)					−0.01	0.00
Pct. vacant					−0.02	0.00
Pct. foreign born					0.00	0.00
Pct. owner occupied					−0.01	0.00
Population density					0.00	0.00
(Intercept)	−0.87	0.03	−0.98	0.02	0.15	0.18
San Jose						
Pct. non-Latino black	−0.01	0.03			−0.01	0.03
Pct. Latino	0.02	0.00			0.03	0.01
Pct. in poverty			0.04	0.01	0.02	0.02
Pct. college educated (ages 25+)					0.02	0.01
Pct. vacant					0.02	0.03
Pct. foreign born					0.01	0.01
Pct. owner occupied					0.00	0.01
Population density					−0.03	0.02
(Intercept)	−0.62	0.22	−0.53	0.12	−2.27	0.98
Philadelphia						
Pct. non-Latino black	0.01	0.00			0.00	0.00
Pct. Latino	0.02	0.00			0.00	0.00
Pct. in poverty			0.05	0.00	0.02	0.00
Pct. college educated (ages 25+)					0.00	0.00
Pct. vacant					0.05	0.00
Pct. foreign born					−0.02	0.00
Pct. owner occupied					0.00	0.00
Population density					0.02	0.00
(Intercept)	−0.95	0.07	−1.19	0.06	−1.29	0.21
Detroit						
Pct. non-Latino black	0.01	0.00			0.00	0.01
Pct. Latino	0.02	0.00			0.01	0.00
Pct. in poverty			0.03	0.00	0.01	0.00
Pct. college educated (ages 25+)					−0.01	0.01
Pct. vacant					0.04	0.00
Pct. foreign born					0.00	0.01
Pct. owner occupied					0.01	0.00
Population density					−0.03	0.02
(Intercept)	−0.73	0.18	−1.41	0.07	−1.26	0.88

change in the neighborhood disorder scale associated with each one-unit difference in the independent variables. The models were all weighted to reflect the population of adults aged 65 and older in each city.

The first panel of Table 2 reports the parameter estimates and standard errors of sequential models of New York City tracts. Living in neighborhoods with a 10 percent higher percentage of black residents was associated with a 0.08 standard deviation increase in physical disorder.[6] A stronger correlation existed between the percentage of Latino residents and disorder; a neighborhood with 10 percent more Latinos had a quarter of a standard deviation more disorder. Estimates of model 2, which includes only poverty, showed a strong association between *neighborhood* poverty and exposure to disorder among older New Yorkers. Ten percent higher poverty was associated with almost half of a standard deviation more disorder.

Adding the covariates described above reduced all three associations. The association with black racial composition virtually disappeared while that of Latino composition was cut by three-quarters. Each 10 percent higher Latino composition was only associated with a 0.06 standard deviation increase in disorder. The influence of the poverty rate after adjusting for covariates was also three-quarters of the unadjusted estimate. The covariates for educational attainment, home ownership rate, vacancy rate, and population density all associated with disorder in the expected directions. Neighborhoods with more foreign-born residents had modestly higher levels of disorder.

The second panel of Table 2 shows that the parameter estimate of black racial composition in San Jose was negative but imprecise. Elders living in neighborhoods with a higher percentage of black residents experienced *less* disorder than those with lower percentages of black residents, but there was a large degree of variation. Elders who lived in neighborhoods with more Latinos were exposed to more disorder. A neighborhood where Latinos made up 10 percent more of the population had 0.20 of a standard deviation more disorder. The correlation was *stronger* after adding covariates. We estimated that the association between poverty rates and disorder in San Jose was similar to that in New York: a 10 percent higher poverty rate was associated with an almost half of a standard deviation higher level of disorder. Adding covariates halved the correlation between poverty rates and disorder and reduced the precision of the estimate.

The third panel of Table 2 shows that race was strongly associated with levels of disorder among Philadelphia neighborhoods that did not control for covariates. Disorder was a tenth of a standard deviation higher for every 10 percent more blacks who lived in a neighborhood and two-tenths higher for every 10 percent more Latinos. Controlling for other neighborhood characteristics eliminated the association between disorder and black racial composition and virtually eliminated the association between disorder and Latino racial composition. Consistent with both New York and San Jose, the unadjusted association between disorder and poverty was about half a standard deviation for each 10 percent larger poverty rate. Adding covariates halved the association between poverty rates and disorder, still leaving a substantial association. The estimates showed positive and precise correlations between vacancy rates and population density and a negative, precise correlation with the percentage of foreign-born residents.

The final panel of Table 2 reports the parameter estimates of coefficients and standard errors for older Detroiters. Black racial composition had a negligible association with physical disorder while Latino racial composition had a more substantial association. A difference of 10 percent in the Latino composition was associated with a 0.20 standard deviation higher level of disorder. A 10 percent increase in the poverty rate was associated with a 0.32 standard deviation increase in disorder. On a standardized basis, this reflected a smaller association than in the other cities. Controlling for other neighborhood characteristics halved the magnitude of the association between disorder and Latino racial composition; controls reduced the association between disorder and the poverty rate by a third. The vacancy rate correlated strongly and positively with the level of disorder. The remainder of the controls did not exert much of an influence, and most were relatively imprecisely estimated.

Between-city variation

Even this small set of four cities revealed that variance across cities in the associations between neighborhood physical disorder and other neighborhood characteristics exists. To show the variation more clearly, we plotted the point estimates and 95 percent confidence bounds for associations in Figure 5. We plotted the associations of percent black, percent Latino, and the poverty rate with disorder for each of the four cities for the unadjusted and adjusted models. Staying consistent with the section above, we plotted the estimates and error bounds associated with a 10 percent increase in each measure.

Comparing the top panel of Figure 5 to the bottom, the variation in the size of the association sizes increased for all three neighborhood factors. Although, one should not make too much of a sample of only four cities, the increasing variation in the point estimates suggests a city-specific pattern of association between neighborhood conditions and racial and economic disparities.

Lessons for policy-makers and researchers

The data contained in Street View images make it possible to increase our study of the variation in neighborhood effects across cities. Understanding the mechanisms that link neighborhoods to individual and collective outcomes requires that we capture the heterogeneity of neighborhoods and neighborhood associations across a variety of neighborhood conditions (Small and Feldman 2012). Measuring neighborhoods and their residents across cities can help us to decompose the influence on racial and economic disparities into different geographic scales. This decomposition should also help policy-makers. Before city governments implement one-size-fits-all policies regarding neighborhood environments, policy-makers should consider the unique variation within their own cities. Having tools readily available can lower the barriers to entry for researchers, especially nonacademic users, who might not have the specialized expertise across geographic, social, and computer sciences necessary to investigate issues of societal concern. We hope that CANVAS provides a step in this direction, but more research is necessary to completely develop systems that promote and disseminate research.

FIGURE 5
Unadjusted (Left) and Adjusted (Right) Beta Coefficients and Confidence Interval Measuring Influence of 10 Percent Change in Percent Black, Percent Latino, and Poverty Rate on Level of Physical Disorder Observed Using Google Street View

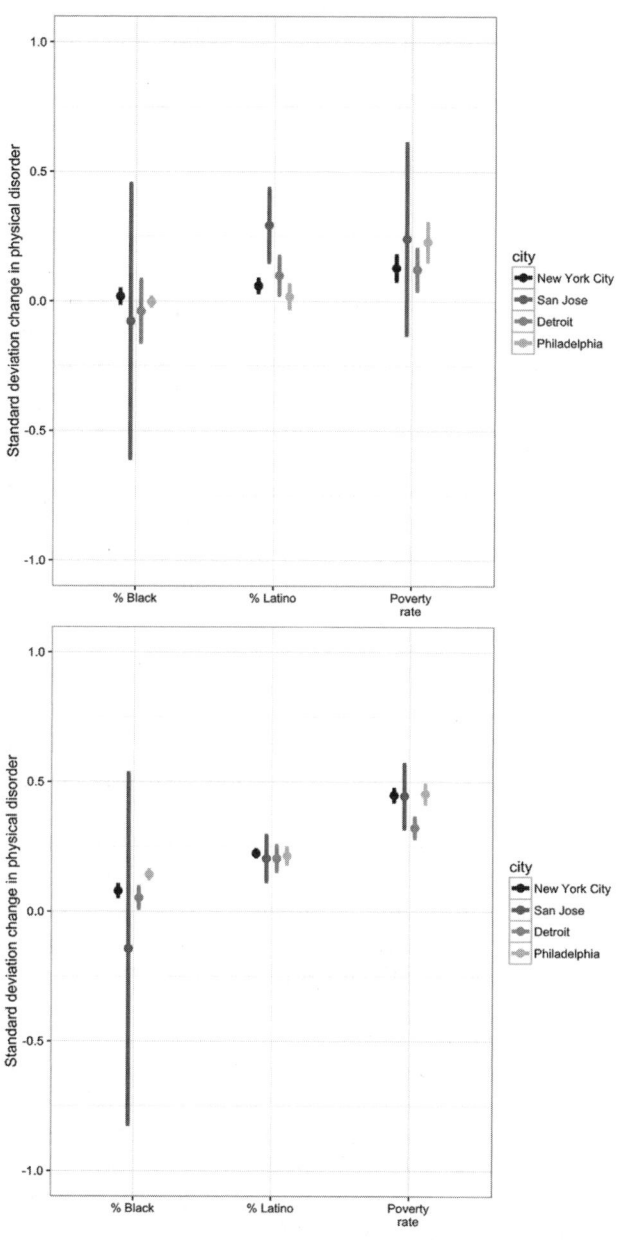

Perils of Using Virtual Neighborhood Audits

Virtual audits using Google Street View make analyses such as those presented above possible. Indeed, Street View makes it conceivable to conduct studies of neighborhood environments not just in a handful of cities, but across the entire nation and even between nations. This possibility provides a new potential resource to understand neighborhood exposures and their associations with individual and collective outcomes. At the same time, however, the promise of virtual audits comes with perils. While some of these perils are not unique to the source of data, other perils arise specifically because of the data source.

Privacy

One major concern that arises by using Street View—indeed when using any online service provider—involves the protection of personally identifiable information of study subjects. Unlike in-person audits, where researchers may observe a subject's specific location, Street View cannot be used to study specific locations without seriously considering the confidentiality of data. Entering an address into Google Map's interface to pull up Street View imagery passes that personally identifiable information to Google, which is a corporate entity not covered by an institutional review board (Bader, Mooney, and Rundle 2016). As a consequence of not being able to enter respondents' addresses into Street View, specific individuals' most proximate physical environments cannot be studied.

Proprietary data

The proprietary control that Google can exert over the underlying image data presents another problem for research. Researchers are subject to Google's terms of service, which Google can, at any time, modify. Future changes to the terms of service might prevent this kind of research or introduce additional costs that make data collection less feasible. In addition, because the data are proprietary, they do not include certain valuable elements of metadata. In particular, we do not know the exact date and time at which the data were collected (only the month and year). Google does not release these data, possibly because revealing them would reveal proprietary information that would help competitors.[7]

Profit motivation

Not only are the data proprietary, but the data are collected for a for-profit corporation interested in maximizing profit to shareholders. The locations at which Google updates imagery will be motivated by changes that drive web traffic to the Street View website. On one hand, this provides an advantage to researchers studying change over time. Rapid change may provide Google incentive to regularly update. On the other hand, updates might be less likely in places where long-term changes occur. As poverty and racial diversity increase, for example, in suburban communities outside of major city centers (Bader and Warkentien 2016; Kneebone and Berube

2015), measuring change might be difficult because updates to the Street View imagery will occur less often. The coverage and frequency of updates might systematically differ between metropolitan and rural settings, as well.

Item mismeasurement

Finally, while Street View can be used to obtain reliable data on a large number of items, other important items tend not to be as reliably or validly measured using Street View. Smells and sounds cannot be evaluated at all. Small items with variation over short periods can be difficult to detect (Bader et al. 2015; Rundle et al. 2011). This makes the measurement of items such as drug paraphernalia, litter, and broken glass unreliable. Because images are taken from the street on top of a car at most locations, measures related to sidewalks can be difficult to assess. Parked cars or shrubbery might obscure views, and even where those obstructions do not exist, items such as cracked or discontinuous sidewalks might be difficult to detect. Item-response measurement models can be adapted to overcome some of these limitations by aggregating data and smoothing over measurement error on individual items (Raudenbush and Sampson 1999), but reducing or eliminating error on the items themselves will likely prove to be impossible.

Another source of measurement error comes from rater fatigue. Quality control is essential, especially because low-frequency items that have the potential to heavily influence scales, for example needles and other drug paraphernalia, require a great deal of attention. But devoting so much attention to a monotonous task leads to high levels of burnout and rater attrition. We have experienced this in our own studies. In one study, we hired undergraduate students to rate streets during a regular semester of coursework. Of the four auditors we hired, two burned out. Even among streets that all four auditors observed, the ratings proved to be erratic and unreliable.

Although virtual audits reduce the cost of collecting data, the problems of rater fatigue can reduce cost savings. To reduce the lost savings, some researchers in computer science have begun incorporating iterative processes of machine and human learning to improve the speed and accuracy of collecting data using Google Street View (Hara, Le, and Froehlich 2013; Hara et al. 2014). Building more collaborations between computer and social scientists would increase the feasibility of developing longitudinal, multisite studies of physical environments.

Conclusion

Google Street View offers a great deal of promise to identify and address the mechanisms that connect neighborhoods to individual and collective outcomes. Virtual audits allow researchers to study the influence of neighborhoods across metropolitan contexts and can help researchers to decompose variation across geographic scales. These data can help to answer long-standing questions about how the unequal distribution of neighborhood conditions might create inequality in society. As we have shown, black and Latino older residents in four cities live in neighborhoods that expose them to higher levels of disorder than white residents. The same is generally true for poor older residents. The size of disparities, however, varies across the four cities. By increasing the scale of data collection, we can increase knowledge on these variations. As we do, it will be important to avoid or address the perils of research using virtual audits and to complement these data with new methods to overcome the limitations of virtual audits.

Notes

1. Google Street View is a registered trademark of Google Inc.
2. The SAT is an example of a psychometric scale that combines answers on individual items into a single scale to (imperfectly) estimate "scholastic aptitude."
3. The 1995 observations were gathered using videotaping equipment inside of vehicles that drove through neighborhoods (Carter et al. 1995; Sampson and Raudenbush 1999).
4. Some segments required manual adjustment because they fell on highways or other segment types that were not designed for uses where residents would be out and about (Mooney et al. 2014).
5. In each city, we fit a variogram based on an exponential spatial decay function to determine the spatial covariance structure that was used to interpolate the values on the 300-foot grid. For more details, see Mooney et al. (2014).
6. Estimates in Table 2 reflect change correlated with 1 percent change rounded to nearest 0.01 standard deviation; estimates presented in text represent values multiplied by unrounded estimate.
7. Google hires contractors to drive the Google Street View car (Douglas 2009) and guards details about how and when data are collected.

References

adaptlab. 2012. Google Street View Project. Available from http://adaptlab.org/gallery/104-2/.
Anguelov, Dragomir, Carole Dulong, Daniel Filip, Christian Frueh, Stéphane Lafon, Richard Lyon, Abhijit Ogale, Luc Vincent, and Josh Weaver. 2010. Google Street View: Capturing the world at street level. *Computer* 43 (6): 32–38.
Bader, Michael D. M., and Jennifer A. Ailshire. 2014. Creating measures of theoretically relevant neighborhood attributes at multiple spatial scales. *Sociological Methodology* doi:0081175013516749.
Bader, Michael D. M., Stephen J. Mooney, Yeon Jin Lee, Daniel Sheehan, Kathryn M. Neckerman, Andrew G. Rundle, and Julien O. Teitler. 2015. Development and Deployment of the Computer Assisted Neighborhood Visual Assessment System (CANVAS) to measure health-related neighborhood conditions. *Health & Place* 31:163–72.
Bader, Michael D. M., Stephen J. Mooney, and Andrew G. Rundle. 2016. Protecting personally identifiable information when using online geographic tools for public health research. *American Journal of Public Health* 106 (2): 206–8.
Bader, Michael D. M., and Siri Warkentien. 2016. The fragmented evolution of racial integration since the civil rights movement. *Sociological Science* 3:135–66.
Bayer, Ada-Helen, and Leon Harper. 2000. *Fixing to stay: A national survey of housing and home modification issues*. Washington, DC: AARP. Available from http://assets.aarp.org/rgcenter/il/home_mod.pdf.
Browning, Christopher R., Reginald A. Byron, Catherine A. Calder, Lauren J. Krivo, Mei-Po Kwan, Jae Yong Lee, and Ruth D. Peterson. 2010. Commercial density, residential concentration, and crime: Land use patterns and violence in neighborhood context. *Journal of Research in Crime and Delinquency* 47 (3): 329–57.
Carter, Woody, Jody Dougherty, and Karen Grigorian. 1995. Appendix III: NORC narrative, videotaping neighborhoods. In *Project on human development in Chicago neighborhoods (PHDCN): Systematic social observation, inter-university consortium for political and social research study 13578*, eds.

Felton J. Earls, Stephen W. Raudenbush, Albert J. Reiss Jr., and Robert J. Sampson. Ann Arbor, MI: Inter-University Consortium for Political; Social Research (distributor).

Chetty, Raj, Nathaniel Hendren, and Lawrence Katz. 2016. The effects of exposure to better neighborhoods on children: New evidence from the Moving to Opportunity Project. *American Economic Review* 106 (4): 855–902.

Clarke, Philippa, and Els R. Nieuwenhuijsen. 2009. Environments for healthy ageing: A critical review. *Maturitas* 64 (1): 14–19.

Clarke, Philippa, Jennifer A. Ailshire, Michael Bader, Jeffrey D. Morenoff, and James S. House. 2008. Mobility disability and the urban built environment. *American Journal of Epidemiology* 168 (5): 506–13.

Clarke, Philippa, Jennifer Ailshire, Robert Melendez, Michael Bader, and Jeffrey Morenoff. 2010. Using Google Earth to conduct a neighborhood audit: Reliability of a virtual audit instrument. *Health & Place* 16 (6): 1224–29.

Clifton, Kelly J., Andréa D. Livi Smith, and Daniel Rodriguez. 2007. The development and testing of an audit for the pedestrian environment. *Landscape and Urban Planning* 80 (1–2): 95–110.

Curtis, Andrew, Dominique Duval-Diop, and Jenny Novak. 2010. Identifying spatial patterns of recovery and abandonment in the post-Katrina Holy Cross neighborhood of New Orleans. *Cartography and Geographic Information Science* 37:45–56.

Diez Roux, Ana V. 2003. Residential environments and cardiovascular risk. *Journal of Urban Health* 80 (4): 569–89.

Diez Roux, Ana V., and Christina Mair. 2010. Neighborhoods and Health. *Annals of the New York Academy of Sciences* 1186 (1): 125–45.

Douglas, Paul. 2009. Behind the scenes with Google Street View. *TechRadar*. Available from http://www.techradar.com.

Duncan, Greg J., and Stephen W. Raudenbush. 1999. Assessing the effects of context in studies of child and youth development. *Educational Psychologist* 34 (1): 29–41.

Farber, Nicholas, Douglas Shinkle, Jana Lynott, Wendy Fox-Grage, and Rodney Harrell. 2011. *Aging in place: A state survey of livability policies and practices*. Washington, DC: National Conference of State Legislatures; AARP Public Policy Institute.

Hara, Kotaro, Vicki Le, and Jon Froehlich. 2013. Combining crowdsourcing and Google Street View to identify street-level accessibility problems. In *Proceedings of the SIGCHI Conference on Human Factors in Computing Systems, CHI '13*, 631–40. New York, NY: ACM.

Hara, Kotaro, Jin Sun, Robert Moore, David Jacobs, and Jon Froehlich. 2014. Tohme: Detecting curb ramps in Google Street View using crowdsourcing, computer vision, and machine learning. In *Proceedings of the 27th Annual ACM Symposium on User Interface Software and Technology, UIST '14*, 189–204. New York, NY: ACM.

Hwang, Jackelyn, and Robert J. Sampson. 2014. Divergent pathways of gentrification racial inequality and the social order of renewal in Chicago neighborhoods. *American Sociological Review* 79 (4): 726–51.

Kennan, Teresa A. 2010. *Home and community preferences of the 45+ population*. Washington, DC: AARP. Available from http://assets.aarp.org/rgcenter/general/home-community-services-10.pdf.

Kirtland, Karen A., Dwayne E. Porter, Cheryl Addy, and Barbara E. Ainsworth. 2003. Environmental measures of physical activity supports: Perception versus reality. *American Journal of Preventive Medicine* 24 (4): 323–31.

Klinenberg, Eric. 2003. *Heat wave: A social autopsy of disaster in Chicago*. Chicago, IL: University of Chicago Press.

Kneebone, Elizabeth, and Alan Berube. 2015. *Confronting suburban poverty in America*. Washington, DC: Brookings Institution.

Ludwig, Jens, Greg J. Duncan, Lisa A. Gennetian, Lawrence F. Katz, Ronald C. Kessler, Jeffrey R. Kling, and Lisa Sanbonmatsu. 2012. Neighborhood effects on the long-term well-being of low-income adults. *Science* 337 (6101):1505–10.

Ludwig, Jens, Lisa Sanbonmatsu, Lisa Gennetian, Emma Adam, Greg J. Duncan, Lawrence F. Katz, Ronald C. Kessler, Jeffrey R. Kling, Stacy Tessler Lindau, Robert C. Whitaker, and Thomas W. McDade. 2011. Neighborhoods, obesity, and diabetes—A randomized social experiment. *New England Journal of Medicine* 365 (16): 1509–19.

Marek, Karen Dorman, Frank Stetzer, Scott J. Adams, Lori L. Popejoy, and Marilyn Rantz. 2012. Aging in place versus nursing home care: Comparison of costs to Medicare and Medicaid. *Research in Gerontological Nursing* 5 (2): 123–29.

Mooney, Stephen J., Michael D. M. Bader, Gina S. Lovasi, Kathryn M. Neckerman, Julien O. Teitler, and Andrew G. Rundle. 2014. Validity of an ecometric neighborhood physical disorder measure constructed by virtual street audit. *American Journal of Epidemiology* 180 (6): 626–35.

Mooney, Stephen J., Daniel J. Westreich, and Abdulrahman M. El-Sayed. 2015. Commentary: Epidemiology in the era of big data. *Epidemiology* 26 (3): 390–94.

Mujahid, Mahasin S., Ana V. Diez Roux, Jeffrey D. Morenoff, and Trivellore Raghunathan. 2007. Assessing the measurement properties of neighborhood scales: From psychometrics to ecometrics. *American Journal of Epidemiology* 165 (8): 858–67.

Odgers, Candice L., Avshalom Caspi, Christopher J. Bates, Robert J. Sampson, and Terrie E. Moffitt. 2012. Systematic social observation of children's neighborhoods using Google Street View: A reliable and cost-effective method. *Journal of Child Psychology and Psychiatry* 53 (10): 1009–17.

Perkins, Douglas, and Ralph Taylor. 1996. Ecological assessments of community disorder: Their relationship to fear of crime and theoretical implications. *American Journal of Community Psychology* 24 (1): 63–107.

Perkins, Douglas D., Paul Florin, Richard C. Rich, Abraham Wandersman, and David M. Chavis. 1990. Participation and the social and physical environment of residential blocks: Crime and community context. *American Journal of Community Psychology* 18 (1): 83–115.

Pikora, Terri J., Fiona C. L. Bull, Konrad Jamrozik, Matthew Knuiman, Billie Giles-Corti, and Rob J. Donovan. 2002. Developing a reliable audit instrument to measure the physical environment for physical activity. *American Journal of Preventive Medicine* 23 (3): 187–94.

Raudenbush, Stephen W., and Robert J. Sampson. 1999. Ecometrics: Toward a science of assessing ecological settings, with application to the systematic social observation of neighborhoods. *Sociological Methodology* 29:1–41.

Reiss, Albert J. 1971. Systematic observation of natural social phenomena. *Sociological Methodology* 3: 3–33.

Rubin, Donald B. 1987/2004. *Multiple imputation for nonresponse in surveys*. Hoboken, NJ: Wiley-Interscience.

Rundle, Andrew G., Michael D. M. Bader, Catherine A. Richards, Kathryn M. Neckerman, and Julien O. Teitler. 2011. Using Google Street View to audit neighborhood environments. *American Journal of Preventive Medicine* 40 (1): 94–100.

Sampson, Robert J. 2012. *Great American city: Chicago and the enduring neighborhood effect*. Chicago, IL: University of Chicago Press.

Sampson, Robert J., and Stephen W. Raudenbush. 1999. Systematic social observation of public spaces: A new look at disorder in urban neighborhoods. *American Journal of Sociology* 105 (3): 603–51.

Sampson, Robert J., and Stephen W. Raudenbush. 2004. Seeing disorder: Neighborhood stigma and the social construction of "broken windows." *Social Psychology Quarterly* 67:319–42.

Savitz, Natalya Verbitsky, and Stephen W. Raudenbush. 2009. Exploiting spatial dependence to improve measurement of neighborhood social processes. *Sociological Methodology* 39 (1): 151–83.

Schafer, Joseph L. 2015. Estimation/multiple imputation for mixed categorical and continuous data [software package]. Available from https://rdrr.io/cran/mix/.

Sharkey, Patrick, and Robert J. Sampson. 2010. Destination effects: Residential mobility and trajectories of adolescent violence in a stratified metropolis. *Criminology* 48 (3): 639–81.

Shet, Vinay. 2014. Go back in time with Street View. Available from https://blog.google/products/maps/go-back-in-time-with-street-view/.

Small, Mario L., and Jessica Feldman. 2012. Ethnographic evidence, heterogeneity, and neighbourhood effects after moving to opportunity. In *Neighbourhood effects research: New perspectives*, eds. M. van Ham, D. J. Manley, N. Bailey, L. Simpson, and D. Maclennan, 57–77. Dordrecht, the Netherlands: Springer.

Taylor, Ralph B., Stephen Gottfredson, and Sidney Brower. 1984. Block crime and fear: Defensible space, local social ties, and territorial functioning. *Journal of Research in Crime and Delinquency* 21 (4): 303–31.

Vargo, Jason, Brian Stone, and Karen Glanz. 2012. Google walkability: A new tool for local planning and public health research? *Journal of Physical Activity & Health* 9 (5): 689–97.

Zhu, Xuemei, and Chanam Lee. 2008. Walkability and safety around elementary schools: Economic and ethnic disparities. *American Journal of Preventive Medicine* 34 (4): 282–90.

Understanding Racial Differences in Exposure to Violent Areas: Integrating Survey, Smartphone, and Administrative Data Resources

By
CHRISTOPHER R. BROWNING,
CATHERINE A. CALDER,
JODI L. FORD,
BETHANY BOETTNER,
ANNA L. SMITH,
and
DANA HAYNIE

Emerging evidence indicates that exposure to areas prone to violence may influence youth well-being. We employ smartphone GPS data on a sample of urban youth to examine the extent of, and potential explanations for, racial disparities in these exposures. We use data from the Adolescent Health and Development in Context study, which continuously collects GPS data from the smartphones of participating youth for a week, to analyze exposure to violent areas. We find that exposure varies significantly across days of the week and between youth who reside in the same neighborhood. African American youth are exposed to areas with substantially higher levels of violence. Residing in a disadvantaged neighborhood is significantly associated with exposure to violence and explains some of the racial difference in this outcome, but neighborhood factors are incomplete explanations of the racial disparity. Characteristics of the locations at which youth spend time explain the residual racial disparity in exposure to violent areas. These findings highlight the importance of youth activity spaces, above and beyond their neighborhood environments.

Keywords: activity spaces; exposure to violence; neighborhoods; adolescents; social disparities

Research on exposure to violence (EtV) during adolescence has demonstrated nontrivial effects of EtV on a range of developmental outcomes including mental health (Browning et al. 2013; Fowler et al. 2009), physical health

Christopher R. Browning is a distinguished professor of sociology and an affiliate of the Institute for Population Research at the Ohio State University. His research focuses on neighborhood and activity space influences on health and development, emphasizing the causes and consequences of social processes such as collective efficacy and network dynamics.

Catherine A. Calder is a professor of statistics at the Ohio State University. Her research interests include spatial statistics, Bayesian modeling, and network analysis, with applications in the social, environmental, and health sciences.

Correspondence: browning.90@osu.edu

DOI: 10.1177/0002716216678167

(Sternthal et al. 2010), educational outcomes (Sharkey 2010), externalizing behaviors (Kirk and Hardy 2014; Sharkey et al. 2012), and risk behaviors/delinquency (Fagan, Wright, and Pinchevsky 2015; Sampson, Morenoff, and Raudenbush 2005). Much of this research has emphasized individual exposures through the experience of victimization or direct witnessing of violence either in the home or community. Findings demonstrate that both victimization and witnessing are consequential for subsequent developmental outcomes of youth, and these effects are observed for experiences occurring in the home and community (Margolin and Gordis 2000). A related stream of research has examined the impact of residence in violent neighborhoods on youth well-being, offering evidence that exposure to violent areas is developmentally detrimental (Carroll-Scott et al. 2013; Massey 2004; Sampson, Morenoff, and Raudenbush 2005; Sharkey 2010). Although the effects of residence in a high-violence neighborhood may operate through direct experiences of witnessing or victimization, high aggregate levels of violence—even if not directly witnessed or experienced—may have independent negative effects on adolescents' well-being (Sharkey 2010).

EtV has been offered as a potential explanation for racial disparities in health and well-being among youth. Black youth typically reside in neighborhoods characterized by substantially higher violent crime rates than other race/ethnic groups (Sampson 2012). Chronic exposure to violent neighborhoods, in turn, is likely to result in disparities in consequential stress processes and behavioral development, with implications for life prospects and long-term risk of chronic conditions (Shonkoff et al. 2012). Accordingly, understanding the factors that account for racial differences in exposure to aggregate levels of violence is a necessary step in the larger effort to reduce health and developmental disparities across race/ethnic groups.

Jodi L. Ford is an associate professor and director of the Stress Science Laboratory in the College of Nursing at the Ohio State University. Her research examines the effects of adverse exposures on chronic physiologic stress during adolescence and linkages to subsequent immune function and physical and mental health outcomes.

Bethany Boettner is a research associate in the Institute for Population Research at the Ohio State University. Her recent research examines neighborhood segregation, social climate, and the structure of shared routine activity locations.

Anna L. Smith is a doctoral candidate in statistics at the Ohio State University. Her research focuses on the study of multiple networks and relating network structure to exogenous information.

Dana Haynie is a professor of sociology and director of the Criminal Justice Research Center at the Ohio State University. Her research interests include the influences of adolescent friendship networks on delinquency and how inmate networks affect current behavior and postrelease outcomes.

NOTE: This research was supported by the National Institute on Drug Abuse (R01DA032371), the William T. Grant Foundation, the National Science Foundation DMS-1209161 ("Bayesian Methods for Socio-Spatial Point Patterns and Networks"), and the Ohio State University Institute for Population Research (NICHD P2CHD058484). Thanks to the Ohio State University Criminal Justice Research Center and Emily Shrider for assistance with acquiring and cleaning crime data from the Columbus Division of Police.

Emerging research on the use of urban space calls into question the standard approach to conceptualizing and measuring adverse exposures, particularly the conventional reliance on administratively defined neighborhoods as proxies for relevant exposure spaces. Employing neighborhood violent crime rates as the relevant exposure measure for resident youth ignores the likely substantial variability in everyday exposure, both within and between youth residing in the same neighborhood. To explain variability in adverse exposures adequately, information on the geographic location of routine activities for urban youth is required. To date, with the exception of a small number of studies that have collected data on additional routine activity locations (Basta, Richmond, and Wiebe 2010; Wiebe et al. 2016; Wikström et al. 2012), rich information on everyday geographic exposures among youth has been unavailable.

The Adolescent Health and Development in Context (AHDC) study includes precise and continuous location data from the smartphones of participating youth over one week; we use AHDC data here to study the geographic exposures of a sample of urban youth to violent areas. We address a number of questions regarding the sources of variability in exposure to aggregate levels of violence. First, we estimate the percentage of variability in exposure to violence in youths' *activity spaces* that rests at the residential neighborhood level (testing the implicit assumption of extant research relying on neighborhood violence rates to estimate the consequences of exposure to violence). Second, we ask whether evidence of racial disparities in residential neighborhood levels of violence extends to the actual routine activity locations of urban youth. Third, we examine a number of hypotheses for observed racial disparities in exposure to activity space violence, asking whether disparities are due to differences in family and residential neighborhood disadvantages or due to constraints beyond these factors that shape the disadvantage of routine activity locations that black youth typically experience.

Exposure to Violent Areas and Urban Youth Outcomes

Exposure to community violence has been linked with a wide range of developmental outcomes among youth. Here we define community EtV to include both direct violent victimization and witnessing of violence, as well as indirect exposure through "hearing about" violence levels or trends within an area to which the youth is regularly exposed. Research has offered mounting evidence of associations between community EtV and behavioral, mental, and physical health outcomes, suggesting that EtV has wide-ranging negative impacts on children (Bingenheimer, Brennan, and Earls 2005; Foster and Brooks-Gunn 2009; Kupersmidt, Shahinfar, and Voegler-Lee 2002; Margolin and Gordis 2004; H. Osofsky and J. Osofsky 2004; J. Osofsky 1999). These associations also appear to be durable. Ford and Browning (2014), for instance, found that males who witnessed violence with a weapon (a shooting or stabbing) and females who were victims of violence (shot, cut, stabbed, or had a knife or gun drawn on them) during adolescence were more likely to have hypertension in young adulthood.

A number of mechanisms have been advanced to explain the link between EtV and compromised health and development. We consider both stress and social learning processes as critical conduits of the adverse impact of EtV. Stress process explanations focus on the role that exposure to severe violence plays in health-consequential physiological stress and socioemotional pathways that have implications for development. Research on the acute effects of exposure to homicide, for instance, demonstrates that, among African American youth, residing in a census block group in which a recent homicide has occurred substantially decreases performance on cognitive evaluations (Sharkey 2010). Sharkey and colleagues point to the role of emotional responses, such as fear and helplessness, in response to residentially proximate homicide events in inducing PTSD-related symptomatology, such as compromised sleep, concentration problems, and increased anxiety.

Acute stress effects of EtV may also be embedded in ongoing chronic stress responses due to routine exposure to high levels of violence. Life course and stress theories (e.g. allostatic load; Elder 1998; Halfon et al. 2013; McEwen and Stellar 1993; Pearlin et al. 2005; Shonkoff et al. 2012; Theall et al. 2013) posit that chronic psychological stress associated with adverse exposures cumulatively affects health through damage to organs and tissues as a consequence of recurrent and prolonged activation of the stress response systems. These negative effects may be more salient if exposure occurs during sensitive developmental periods or life transitions, such as adolescence (Elder, Johnson, and Crosnoe 2003; Halfon et al. 2013). Recent research supports this hypothesis, as youth living in neighborhoods characterized by violence, poverty, or "stressors" (e.g., composites of poverty, disorder, and/or violence) were found to have irregular salivary diurnal patterns or acute stress reactivity (Brenner et al. 2013; Chen and Paterson 2006; Hackman et al. 2012; Rudolph et al. 2014), and in one study, elevated levels of hair cortisol (Vliegenthart et al. 2016). Although the detrimental effect of exposure to high-violence neighborhoods is likely to work, in part, through direct victimization and witnessing experiences, the impact of routine exposure to the threats posed by high-violence areas may independently affect well-being.

Additional pathways through which exposure to violence may operate include the role of social learning and informal social control processes in influencing behavioral proclivities among youth. In a well-known ethnographic study of an urban, African American neighborhood, Anderson points to the role of violence as a defensive posture. Residing in neighborhoods observed or expected to be characterized by high levels of violence is hypothesized to propagate such violence by leading youth to maintain violent demeanors or to act violently to reduce the subsequent threat of victimization (Anderson 1999). In this view, violence becomes a self-reinforcing adaptation to routine exposure to threatening environments. High levels of neighborhood violence may also cue the breakdown of social order, leading to reduced guardianship of public space (Felson and Cohen 1980) and expectations among youth that prosocial norms will not be as effectively enforced (Skogan 1990). Existing evidence indicates that residing in high-violence neighborhoods is linked with increased risk of individual-level violence (Sampson, Morenoff, and Raudenbush 2005) and aggression (McMahon et al. 2013; Vanfossen et al. 2010), pointing to the potential relevance of learning and informal social control processes in the link between neighborhood violence and behavioral outcomes (although few studies have attempted to uncover the mechanisms responsible for neighborhood violence effects). In short, evidence suggests that exposure to high aggregate levels of violence is harmful across a number of domains for urban youth and may account for a nontrivial proportion of racial disparities in these outcomes.

Neighborhoods and Activity Spaces as Exposure Spaces for Youth

Research on "neighborhood effects" expanded after the publication of William Julius Wilson's *The Truly Disadvantaged* (1987). By highlighting the spatial concentration of joblessness, teenage pregnancy and childbirth, delinquency and crime, and other phenomena, Wilson helped to spur a sustained interest in understanding community-level factors that shape the life prospects of the urban poor. The subsequent decade of the 1990s saw the emergence of multilevel research designs that disentangled individual and contextual contributions to a range of outcomes. On balance, this research effort led to acknowledgment of "neighborhood effects" on developmental trajectories and the importance of context in youths' lives (Sampson, Morenoff, and Gannon-Rowley 2002).

Although a fruitful approach to understanding and measuring contextual effects, the multilevel design that embeds individuals in their residential neighborhood—linking features of the latter with outcomes—is potentially problematic. Most notably, the approach assumes that the residential census unit employed (typically tract or block group) captures the relevant exposure space. Indeed, the observation that individual mobility extends beyond the boundaries of residential neighborhoods is almost as old as urban sociology itself (Browning and Soller 2014; McKenzie 1921). The practice of embedding youth in residential census units nevertheless became the default design for neighborhood effects studies due, in large part, to limitations in available data on exposures. A limited number of studies (Cook et al. 2002; Massey and Brodmann 2014) incorporated information on both neighborhood and school contexts (Bowen and Bowen 1999; Kirk 2009; Owens 2010; Teitler and Weiss 2000). These "multicontextual" approaches, however, remained rare through the 1990s and 2000s.

More recently, an emerging body of research has employed an "activity space" approach to assessing the contexts of exposure. Activity spaces capture all the locations to which individuals are regularly exposed (Browning and Soller 2014; Graif, Gladfelter, and Matthews 2014; Jones and Pebley 2014; Kwan 2009, 2013; Sharp, Denney, and Kimbro 2015; Wikström et al. 2012). In this view, theories of contextual impact that rely on an exposure mechanism should empirically determine the span of adolescent exposures to produce an unbiased estimate of the impact of context. Wikström et al. (2012), for instance, employ a space-time budget approach adapted from geographic methods (Hägerstrand 1970) to capture locations, activities, behaviors, and social network partner presence as they vary over a four-day period. The ability to situate acts of delinquency and crime in space and time allows for more precise assessment of context effects than conventional approaches that rely on residential neighborhood units of unknown relevance with respect to exposure.

The location and characteristics of residential neighborhoods are likely to be important determinants of activity space features. In some instances, residential neighborhoods may, in fact, overlap significantly with activity space locations. However, we argue that the daily travel trajectories of urban youth are likely to extend beyond the residential boundary, resulting in both substantial within-individual variability in everyday exposures and variability between individual youth who reside in the same neighborhood. Observed heterogeneity in everyday exposure to areas characterized by high levels of violence among youth who reside in the same neighborhood would indicate that the conventional neighborhood effects design is likely to mischaracterize exposures, attenuating their estimated impact on well-being. Moreover, emphasis on residential neighborhood violence rates as the relevant exposure shifts research questions on the origins of differential exposure to residential mobility patterns (i.e., determinants of neighborhood in which to reside) rather than everyday routine activity patterns (i.e., determinants of areas in which time is spent). Our approach focuses on estimating and explaining variability in exposure to violent areas in the course of everyday routines, emphasizing the magnitude and determinants of racial disparities in this outcome.

Explaining Variability in Activity Space Violence

Extant research on racial disparities in health and well-being among urban youth has emphasized the role of family resources, family processes, and neighborhood-level determinants (Wickrama, Noh, and Bryant 2005; Zimmerman and Messner 2013). Here, we apply these approaches to understanding the allocation of time across nonhome, routine activity locations characterized by varying levels of violent crime. First, the *family investment model* highlights the role of family socioeconomic status (SES) in providing opportunities for investment in children that promote health and development. Limitations on economic resources constrain families to focus on more immediate material needs (Mayer 1997) to the potential detriment of developmentally beneficial exposures. Parental investments include provision of an adequate standard of living, promoting educational and extracurricular opportunities, and engaging with youth on developmentally significant tasks such as academic advancement (Conger and Donnellan 2007; Simons et al. 2016). A key component of family investment models is parents' choice of neighborhood, as the residential context is seen as a critical source of both risk and resources for youth. Presumably, the quality of the neighborhood context is relevant to the extent that it captures actual exposures that contribute to youth outcomes. Accordingly, an extension of the family investment model would take into account characteristics of youths' nonhome activity locations to more directly capture parental investments as they relate to youths' actual exposures (moving beyond the assumption that residential neighborhood characteristics serve as a proxy for these exposures). Racial disparities in youth exposure to violent activity space locations would then be hypothesized to follow from differences in family socioeconomic resources and constraints—for example, income, education, household size, family structure (capturing the availability of parental time for investment in youth), and recent financial problems.

Second, like the family investment model, the *family stress model* views socioeconomic factors as critical determinants of the well-being of children. However, stress processes and compromised parenting flowing from economic distress are seen to play a critical role in mediating the impact of SES (Conger et al. 1994). In this view, financial problems adversely impact caregiver stress and relational conflict among parents. In turn, parent-child interactional dynamics may be negatively affected, resulting in reduced warmth and support, and disruptions in monitoring and supervision of youth activities and behavior. The more frequent occurrence of economic stress among African American families, in this model, would be reflected in compromised family social support and informal social control capacity, with implications for youths' exposure to higher risk activities and locations.

Third, an alternative approach to understanding racial disparities in high-risk exposures shifts attention away from family-level resources and processes toward the *neighborhood* context in which African American families typically reside by comparison with other race/ethnic groups (Pattillo-McCoy 1999). African American youth are more likely to reside in racially segregated or heterogeneous neighborhoods with lower SES and higher levels of population turnover, all of which have been linked with higher rates of violent crime (Kubrin and Weitzer 2003; Sampson, Morenoff, and Raudenbush 2005; Sampson, Raudenbush, and Earls 1997). Thus nonhome exposures may be constrained by the availability of spatially proximate lower violent crime areas in which to engage in everyday routine activities. Neighborhood effects approaches to explaining racial disparities in exposure to violence have relied heavily on this idea (Buka et al. 2001; Selner-O'Hagan et al. 1998) and have assumed that family-level factors alone are unlikely to completely explain risk of exposure to violence, given the potential for disadvantage in the residential environment to constrain access to less violent activity locations.

Finally, the *activity space* approach acknowledges the likely substantial variation in the location and characteristics of routine activity settings. Neighborhood approaches risk a form of determinism by linking residential context characteristics with features of activity locations, absent recognition of the complexity of everyday activity spaces and the choices and constraints urban actors face in navigating their environments. Although neighborhoods are likely to be important determinants of activity location characteristics, urban families may encounter opportunities for beneficial exposures (e.g., informal, nonpurposive interactions that shape access to higher quality institutional and activity options [Small 2009]) or constraints on the capacity to avoid higher risk locations. In the latter case, even African American families with relatively high levels of advantage residing in higher SES neighborhoods may maintain ties to higher risk areas through relatives, friends, or institutional involvements. For older, more mobile youth, residence in a more advantaged community may still be accompanied by time spent in higher risk areas if peers are concentrated in such spaces (and are less prevalent within the home neighborhood [Clampet-Lundquist and Massey 2008]). The social benefits accrued through these activities may be willingly sought; nevertheless, African American families face such trade-offs at a far higher rate than other groups, potentially resulting in observed racial disparities in exposure to violent areas even after accounting for family- and neighborhood-level factors.

In summary, we employ smartphone-based global positioning system (GPS) data on the travel paths of a sample of youth to estimate variability in the everyday level of exposure to violent crime in a relatively large urban area. We decompose this variation within an individual (across days of a week), between individuals within residential neighborhoods, and across neighborhoods. We then consider family, neighborhood, and activity space determinants of time-weighted exposure to violent areas employing three-level, random effects linear models of this outcome. These analyses offer the first large-scale investigation of exposure to violent crime in the activity spaces of urban youth.

Baseline Data and Measures: The Adolescent Health and Development in Context Study

The Adolescent Health and Development in Context (AHDC) study is a longitudinal, representative data collection effort that focuses on multiple contexts of youth development. The study area is a contiguous space within Interstate 270 (including a majority of Columbus, Ohio, and a number of suburban municipalities). The study also includes an oversample of predominantly African American, low-income census tracts.[1] For the purposes of the current analysis, we use data on youth who reside within both the study area and the City of Columbus to align with available data on violent crime (limited to Columbus). The resulting sample includes a substantial number of African American youth, allowing for effective estimation of racial disparities in violent exposures.

The study design involves multimodal data collection over the course of a week. First, an entrance survey with a focal youth and his/her caregiver is conducted. Caregiver interviews cover a range of topics, including demographic and socioeconomic background, household composition, family structure and marital status, employment and income, health, social support, and alcohol/substance use. Caregivers also report youth characteristics, including behavior, mental/physical health, schooling, family conflict, and legal troubles.

Following the entrance survey, a seven-day smartphone-based GPS tracking and ecological momentary assessment (EMA) data collection period occurs (GPS week). The EMA involves five daily mini surveys, including questions on location, activity, and assessments of the immediate spatial context. The continuous

GPS tracking feature of the AHDC allows for collection of real-time data on the locations at which youth spend time. The phone app prioritizes spatial data from more accurate GPS satellites and saves location data every 30 seconds when connected. If no GPS satellite position has been saved in the past 10 minutes, location coordinates based on cell tower network position are collected every 60 seconds. We collect battery level of the phone to assess missing GPS time periods due to low battery. The GPS data are uploaded to secure servers every hour.

Finally, an exit survey space/time budget is administered to the youth, during which the interviewer walks the youth through five of the seven days covered by the GPS week—the two most recent weekdays and Friday, Saturday, and Sunday. A specialized software application has been designed to employ both GPS inferences on activity locations and EMA data to help the youth recall his or her travel path for the week, facilitating additional collection of information on locations and activities for periods not captured by the EMA (providing continuous data for five days). This process cleans up noise in the GPS coordinates that is present even during stationary periods and provides stable location coordinates for analysis. The raw GPS data are processed through a convex hull detection algorithm to determine stationary periods, which iteratively creates a minimum bounding polygon for a given set of coordinates. When new coordinates increase the size of the convex hull by 100 percent,[2] the algorithm infers that the respondent has left the previous activity location and has begun a travel period. The convex hull can be visualized as the shape formed by a rubber band that has been stretched around the set of points. Convex hull methods have been used recently in several community detection and activity space projects (Chan et al. 2014; Yin et al. 2013). Estimates of GPS data coverage indicate that gaps between logged GPS coordinates greater than five minutes account for only 8 percent of total time for the GPS week.

Data Integration

For the current study, we use survey-based information on demographic and other background characteristics and GPS-based routine activity location data from AHDC youth who reside within the City of Columbus ($N = 878$). We merge those data with (1) geocoded crime statistics from the Columbus Division of Police and (2) American Community Survey data to produce estimates of exposure to violence and other census-based structural characteristics at the block group level.

Measures

We geocode activity space locations from the cleaned GPS data (via the exit survey space/time budget) with valid latitude and longitude to 2010 census block groups. We then generate time-weighted average exposures to violent crime and other census-based covariates for each day, using the number of minutes spent at each location during waking hours as the weighting factor. Adjusting the daily average exposure by time allows for locations with longer associated durations to contribute more to the daily average than locations at which less time was spent. Waking time is reported daily on EMA mini surveys and mean imputed for weekend and weekday separately where self-reports are missing. Note that youth spend some days entirely at home, in which case the youth are not exposed to a nonhome violent crime rate and therefore the day is not included in the analyses (as noted in the missing data section below). If the youth does spend time outside the home on a given day, the day is included in the analyses.

The dependent variable is the time-weighted average of *violent crime*, measured using the violent crime rate (combining aggravated assault, homicide, and robbery) per 1,000 residents in the block group.[3] We geocoded crime reports from 2005 through 2012,[4] provided by the Columbus Division of Police, to 2010 census block group boundaries. The total population measure used as the denominator of the violent crime rate accounts for only the portion of the block group that is within the city of Columbus if the block group is split across municipalities. An activity location is assigned a violent crime rate if it falls within a block group that is partially or wholly inside Columbus. Therefore, the dependent variable does not account for the total amount of time spent outside the home as it excludes time spent outside Columbus. We control for the number of minutes of nonhome time with valid spatial coordinates to account for differences in observed time across days. Although our focus in this analysis is on exposure to aggregate activity space violent crime, bivariate analyses indicate that adolescents with higher average exposure to violent areas report significantly more lifetime exposure to both direct victimization and indirect witnessing of severe violence.

Measures of day-level time-weighted exposure to the following block group characteristics were constructed using 2009–2013 data from the American Community Survey. *Economic disadvantage* is a standardized scale of the following four items: (1) poverty rate, (2) unemployment rate, (3) the percentage of households that are female headed, and (4) the percentage of households that receive cash assistance. *Residential instability* is a standardized scale consisting of the percent of residents ages one and older who have moved in the past year and the percent of occupied housing units that are renter occupied. *Racial diversity* is the sum of the squared proportions of white, Latino, black, Asian, and other race/ethnicity populations in the tract subtracted from 1 ($1 - [\Pr(\text{white})^2 + \Pr(\text{black})^2 + \Pr(\text{Asian})^2 + \Pr(\text{Hispanic})^2 + \Pr(\text{other race})])^2$. Higher values indicate more racially/ethnically diverse activity spaces, as the diversity index captures the likelihood that any two randomly selected individuals are of different race/ethnic groups. We also include a measure of *proportion African American*. At the day level we also control for whether the day is a *weekday* compared to a weekend day.

We include a number of youth and family characteristics that may account for differences in activity space violent crime. We include youth *age* in years and whether the youth is *male* versus female. Race/ethnicity of the youth is represented as four dummy variables indicating *non-Hispanic black*, *Hispanic*, *Asian*, or *multiracial* with *non-Hispanic white* as the omitted reference category. We include whether the caregiver is *foreign born* and whether he/she is *married, cohabiting, single, or other* (married is the reference category). Household level resources are measured by *caregiver education* (ten categories of highest completed, ranging from elementary school to a professional degree), *household size*, and *household income* (fourteen categories ranging from under $10,000 in the previous year to more than $250,000). *Family hardship* is the average level of agreement with three caregiver reports of difficulty paying rent or mortgage, paying utility bills, and putting off buying something they needed like food or medical care because they did not have enough money during the previous 12 months. To assess family supervision and monitoring capacity, youths report on *family control*, a subscale of the Family Environment Scale (Moos and Moos 1994). The number of mostly true responses (vs. mostly false) are counted, so that higher values indicate more family social control. The scale includes nine items, such as "there are very few rules to follow in our family (reverse coded)"; "there are set ways of doing things at home"; "rules are pretty inflexible in our household"; and "you can't get away with much in our family." We also include *family support*, which is a five-item adolescent report indicating the average level of agreement on a three-point scale (Cronbach's α = .81) with the statements such as: "my family lets me know they think I'm a worthwhile (valuable) person"; "people in my family have confidence in me"; and "I know my family will always stand by me" (Turner, Frankel, and Levin 1983). We also include a control for *season* at the youth/family level.

At the neighborhood level we control for a range of characteristics of the residential block group using 2009–2013 American Community Survey data. We include four measures, constructed at the block group level in the same manner as the activity space block group characteristics: neighborhood *economic disadvantage, residential stability, racial diversity*, and *proportion African American*.[5]

Missing data

We start with 4,012 days that have 1,440 minutes (a full day) or less of spatial data coverage from 878 youth who live in the city of Columbus.[6] We remove 138 locations (0.5 percent) with missing or problematic spatial coordinates, which affects 17 days (primarily days with no phone-based GPS due to disabled location services or the phone not being charged). We then remove 727 days[7] where all recorded time was spent at home, during which the youth had no nonhome violent crime exposure. Finally, we limit the analysis to only locations visited within the City of Columbus during waking hours for which we have geocoded violent crime rates available, dropping 182 days and 21 youth with no time spent at nonhome Columbus locations. For days in the analysis, 90 percent of the total waking nonhome time was spent in locations in Columbus. Listwise deletion of cases missing on independent variables results in a sample size of 2,643 days nested with 729 youth[8] and 351 block groups. We employ multiple imputations using Stata's mi impute chained equations commands to address missing data in youth/family-level independent variables for a final sample of 3,086 days nested within 857 youth and 373 block groups.[9]

Analytic Strategy

We employ three-level linear models, nesting average log violent crime exposures at the day level (level 1) within individual youth (level 2) within home block groups (level 3). We define our outcome, Y_{ijk}, to be the average, time-weighted, violent crime rate of the nonhome areas to which youth j, residing in block group k, was exposed on day i. Our analytic approach allows us to decompose the variation in exposure to violent areas across days, individuals, and neighborhoods in unconditional models (without covariates included) and adjust coefficient standard errors for clustering of cases within individual and block groups in conditional models. We begin by decomposing variance in the outcome across levels and constructing intraclass correlations for each level of the nested data structure (model 1), using the following model:

$$Y_{ijk} = \pi_{0jk} + e_{ijk} \qquad e_{ijk} \sim N(0, \sigma^2)$$

$$\pi_{0jk} = \beta_{00k} + r_{0jk} \qquad r_{0jk} \sim N(0, \tau_\pi)$$

$$\beta_{00k} = \gamma_{000} + u_{00k} \qquad u_{00k} \sim N(0, \tau_\beta)$$

In this unconditional model, Y_{ijk} is modeled as a function of π_{0jk}—a youth-specific mean violent crime exposure level over the course of the GPS week—and e_{ijk}: a day-specific deviation from the youth mean exposure (assumed to be independently, normally distributed with variance σ^2). Comparable models are fit to youth and block group mean log violent crime exposures at levels 2 and 3, respectively. Intraclass correlations (ICCs) capture the proportion of the variability at each level of the model in this unconditional specification; e.g., the proportion of the variability at the day level is simply $\sigma^2 / (\sigma^2 + \tau_\pi + \tau_\beta)$.

To test hypotheses regarding any observed racial differences in exposure to violent areas, we fit a series of conditional models, incorporating key covariates at each level. Each of these models is a special case of the following in which all distributional assumptions are identical to the unconditional model introduced above:

$$Y_{ijk} = \pi_{0jk} + \sum_p \pi_{p00} (\text{Activity Space Disadvantage})_{pijk} + e_{ijk}$$

$$\pi_{0jk} = \beta_{00k} + \sum_{q_1} \beta_{q_100} (\text{Race Ethnicity})_{q_1jk} + \sum_{q_2} \beta_{q_200} (\text{Family Disadvantage})_{q_2jk}$$

$$+ \sum_{q_3} \beta_{q_300} (\text{Family Stress})_{q_3jk} + r_{0jk}$$

$$\beta_{00k} = \gamma_{000} + \sum_s \gamma_{s00} (\text{Tract Disadvantage})_{sk} + u_{00k}$$

In conditional models, we begin by entering covariates at level 2, including a measure of race/ethnicity, focusing specifically on estimating the African American–white disparity in exposure to violent areas (controlling for weekend at the day level; model 2). We then turn to family resources and investment capacity (household income, caregiver education, caregiver marital status, and economic distress; model 3) and family stress outcomes (youth-reported family control and family social support; model 4) to determine whether these factors account for the racial disparity in exposure to violent areas. We then include measures of tract-level disadvantage (model 5) at level 3 to examine whether residential characteristics are significantly associated with exposure to violent areas and further explain the racial disparity. Finally, we incorporate activity space disadvantage measures (at level 1) to assess the independent role of these covariates in influencing exposure to violent areas and explaining any residual racial difference.

Results

Table 1 presents descriptive statistics for the day, youth, and neighborhood variables included in the analyses. Of the AHDC youths who reside in the City of Columbus, 35 percent are non-Hispanic white, and 48 percent are non-Hispanic black. We begin by fitting the unconditional model (model 1 of Table 2) to decompose variance in exposure to violent areas across levels of analysis. Under the assumption that neighborhood (in this case, block group) level violence rates are the relevant exposure to capture the contextual impact of aggregate violence, the ICC would result in 100 percent of the variability at the neighborhood level (i.e., youth who reside in the same neighborhood will have the same level of exposure to violence on an everyday basis). In contrast to this expectation, we find that the block group accounts for 35 percent of the variance in exposure to violent areas, with 28 percent at the youth (within neighborhood) level and 37 percent at the day (within individual) level. Consequently, youth experience substantial variability by day in average levels of activity space violence and, among youth who live in the same neighborhood, exposure to violent areas varies nontrivially. These results indicate not only that youth venture beyond the borders of their neighborhood environment in their everyday activities, but also that nonneighborhood environments are heterogeneous with respect to violence rates.

Models 2 through 6 test key hypotheses regarding racial disparities in exposure to violent areas and their explanation. Model 2 incorporates race/ethnicity dummy variables and controls for weekend, season, day-level amount of time spent outside the home, gender, and age. Race/ethnic dummy variables indicate that being black is powerfully positively associated with activity space violence ($p < .001$) as is being multiracial. Specifically, being African American is associated with a 66 percent increase in the average violence level of areas to which youth are exposed outside of the home (time-weighted). Being Latino is marginally positively associated with activity space violence ($p < .10$).

Model 3 adds measures of family resources to examine their unique association with activity space violence and the extent to which these factors explain the observed substantial disparity by race. Among family resource variables, we find that

TABLE 1
Descriptive Statistics of Day-, Youth-, and Neighborhood-Level Variables in the Analysis, from the Adolescent Health and Development in Context Study

	Mean	Standard deviation
Youth characteristics (N = 857)		
Winter	0.21	
Spring	0.20	
Summer	0.27	
Autumn	0.31	
Age	14.19	1.85
Male	0.47	
Non-Hispanic white	0.35	
Non-Hispanic black	0.48	
Hispanic	0.06	
Asian	0.01	
Multiracial	0.10	
Caregiver foreign born	0.05	
Household size	4.68	1.71
Caregiver education	5.63	1.57
Household income	4.94	3.37
Caregiver married	0.46	
Caregiver cohabiting	0.12	
Caregiver single	0.25	
Caregiver other marital status	0.17	
Family hardship	2.22	1.04
Family control	5.34	2.06
Family support	2.62	0.42
Neighborhood (N = 373)		
Proportion black	0.37	0.31
Racial diversity	0.39	0.18
Economic disadvantage	0.43	0.92
Residential instability	0.54	0.91
Daily weighted activity location exposures (N = 3,086)		
Logged violent crime rate	1.70	1.20
Proportion black	0.34	0.28
Racial diversity	0.39	0.15
Economic disadvantage	0.28	0.77
Residential instability	0.59	0.81
Weekend day	0.36	
Minutes of waking nonhome time	563.05	311.04

TABLE 2
Multilevel Linear Models of Daily Time-Weighted Exposure to Violent Crime with Day, Youth, and Neighborhood Predictors

	Model 1	Model 2	Model 3	Model 4	Model 5	Model 6
Youth characteristics						
Season		−0.008	−0.008	−0.007	−0.010	−0.003
Minutes of waking nonhome time		0.000	0.000	0.000	0.000	0.000
Age		0.025	0.033°	0.030	0.028	0.022
Male		−0.071	−0.061	−0.051	−0.050	0.003
Non-Hispanic black		0.664°°°	0.584°°°	0.598°°°	0.330°°°	0.110
Latino		0.234	0.237	0.253	0.110	0.091
Asian		−0.093	−0.079	−0.082	−0.191	−0.262
Multiracial		0.433°°°	0.373°°°	0.377°°°	0.240°	0.205°°
Caregiver foreign born		0.000	−0.053	−0.054	0.034	−0.028
Household size			0.039°	0.038°	0.020	0.005
Caregiver education			−0.031	−0.031	0.003	0.012
Household income			−0.044°°°	−0.042°°°	−0.024°	−0.017
Caregiver cohabiting			0.122	0.117	0.042	0.026
Caregiver single			0.081	0.076	0.023	−0.031
Caregiver other marital status			0.135	0.130	0.091	0.016
Family hardship			−0.010	−0.012	−0.009	−0.005
Family control				−0.008	−0.009	−0.005
Family support				−0.111	−0.109	−0.043
Neighborhood level						
Proportion black					0.605°°°	0.043
Racial diversity					0.465°	0.088
Economic disadvantage					0.274°°°	0.162°°°
Residential instability					0.060	0.020
Daily activity location exposure						
Proportion black						0.985°°°
Racial diversity						1.126°°°
Economic disadvantage						0.305°°°
Residential instability						0.329°°°
Weekend day		−0.011	−0.011	−0.011	−0.013	−0.046
Constant	1.703°°°	1.092°°°	1.210°°°	1.567°°°	1.006°°	0.345
Variance components						
Day	0.767	0.766	0.766	0.766	0.766	0.655
Youth	0.569	0.565	0.577	0.577	0.562	0.414
Home block group	0.725	0.606	0.519	0.516	0.396	0.322

°°°$p < .001$; °°$p < .01$; °$p < .05$;

household size is positively associated with activity space violence ($p < .05$), while household income is negatively associated with the outcome ($p < .01$). The addition of family resource variables reduces the coefficient for African Americans by 12 percent and the coefficient for multiracial identification by 14 percent.

Model 4 considers the association between measures of family control and social support as potential social process outcomes of economic stress that may influence activity space disadvantage. Neither measure was a significant predictor of activity space violence, nor were coefficients for race/ethnicity altered in this model specification.

Models 5 and 6 include measures of neighborhood and activity space disadvantage, respectively. Consistent with prior research, youth who reside in more segregated neighborhoods with larger proportions of African American residents are exposed to more violent areas. Similarly, higher levels of economic disadvantage and race/ethnic diversity in the residential block group are also positively associated with time spent in violent areas. Only residential instability did not achieve significance in this model specification. Of note are declines in the magnitude of individual-level race coefficients. Coefficients for African American, Latino, and multiracial identification all decline by roughly 50 percent in model 5 compared with baseline estimates in model 2, but the coefficients for African American and multiracial remain statistically significant. In the former case, the effect remains highly statistically significant ($p < .001$) and nontrivial: being African American is associated with a 33 percent increase in the average violence level of nonhome exposures.

Finally, model 6 adds day-level measures of activity space disadvantage (time-weighted). Unsurprisingly, we find that characteristics of the activity spaces youth encounter are associated with the violence rates of those areas. All four measures of activity space disadvantage are powerfully positively associated with the outcome ($p < .001$). Moreover, inclusion of these measures results in an 83 percent reduction in the magnitude of the baseline coefficient for African Americans in model 2; the coefficient is no longer statistically significant in model 6. With the exception of home block economic disadvantage (which remains a significant predictor of activity space disadvantage), coefficients for residential neighborhood disadvantage are also no longer significant in model 6. Of note is the effect of multiracial identity, which remains a significant predictor of activity space violence ($p < .05$) after taking into account neighborhood and activity space characteristics.

Figure 1 presents the expected values of violent crime rate by race/ethnicity for models 2 through 6, holding all other covariates at their means. The expected logged crime rates are exponentiated to present rates in their original metric. As noted in the discussion above, the gap between African American and white youth narrows substantially with the addition of residential disadvantage in model 5 and activity space disadvantage in model 6.

Conclusion

Our analyses offer the first examination of violence levels in the actual GPS-measured activity spaces of urban youth. Exposure to violence is a well-established risk

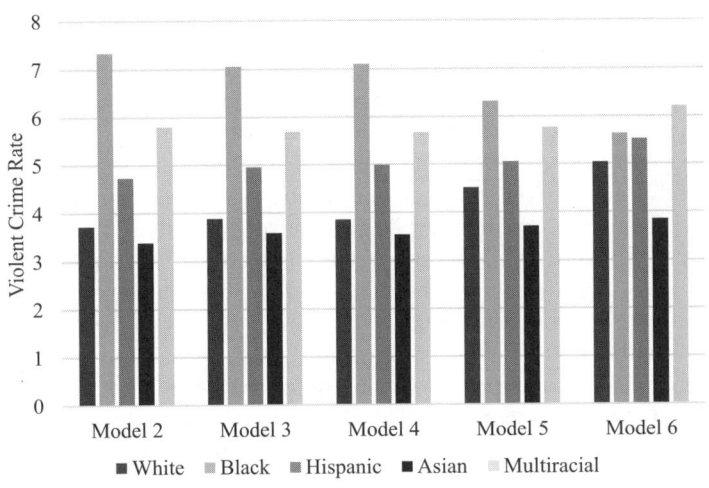

FIGURE 1
Predicted Daily Activity Space Violent Crime Rates by Race/Ethnicity

NOTE: Predicted log violent crime rates are exponentiated.

associated with residing in large urban areas. Previous studies have documented that exposure to violence has negative developmental consequences, and evidence suggests that exposure to violent neighborhoods may be detrimental independent of the experience of direct victimization (Sharkey 2010), pointing to the importance of understanding racial disparities in exposure to violence.

With detailed information on the travel paths of a sample of youth over the course of a week, we found that a majority of the variability in exposure to violent areas is at the day and youth level rather than the residential neighborhood level. This finding points to the substantial heterogeneity in violence exposures that urban youth experience. Employing residential neighborhood violence rates alone significantly misrepresents this heterogeneity.

We next examined hypotheses regarding the sources of variability in exposures to violent areas, considering family, neighborhood, and activity space determinants. Although family resources, specifically income and household size, were associated with exposure to violent areas, they accounted for a relatively modest proportion of the observed racial disparity. Family process measures were not associated with the outcome. In contrast, neighborhood disadvantage measures accounted for roughly 39 percent of the baseline African American–white disparity in exposure to violence areas. Measures of segregation, race/ethnic heterogeneity, and socioeconomic disadvantage in the home block group were powerful predictors of activity space disadvantage, supporting the hypothesis that residential neighborhoods shape activity space exposures to a substantial extent.

Nevertheless, neighborhood measures only partially accounted for the racial disparity in exposure to violent areas. Indeed, the incorporation of neighborhood disadvantage and family-level predictors explained only half the baseline racial disparity.

Even after including these measures, the race effect was statistically significant and nontrivial. Including measures of activity space disadvantage explained an additional 33 percent of the baseline racial disparity and rendered the African American–white coefficient insignificant. Considered in the absence of neighborhood covariates, incorporation of activity space disadvantage measures explains roughly 70 percent of the racial disparity (with an insignificant residual racial disparity).

Although we would expect that characteristics of activity space locations would be associated with the violent crime rates of those areas (just as a long history of neighborhood research on violence has demonstrated), the proportion of the racial disparity that is accounted for by considering activity space locations versus residential neighborhoods alone speaks to the necessity of looking beyond the neighborhood environment to understand the everyday exposures of urban youth. The substantial heterogeneity in exposures observed over time and within residential neighborhood further supports this assertion.

Beyond the analysis of exposure to violent areas, these findings raise additional concerns regarding the continued reliance on conventional "neighborhood effects" approaches to assessing contextual effects on youth development. First, the literature examining neighborhood influence on child and adolescent outcomes is now vast and is expanding. Yet the practice of investigating neighborhood effects by embedding youth in their residential unit (and relating characteristics of this unit to an outcome) is likely to significantly misrepresent actual developmentally relevant exposures. The ongoing controversy regarding the magnitude and import of neighborhood effects is based on extant research that largely employs a potentially crude characterization of nonhome exposure. The current dominant research design, then, may result in attenuated estimates of contextual influence due to inadequate measurement of exposures.

Second, a focus on activity spaces points to the need to understand how such spaces are shaped in the context of everyday routines. Clearly, the problem of selection (notoriously afflicting neighborhood research) remains a significant concern in this approach. Choices regarding routine activities and their locations are shaped by a complex nexus of individual, social network, and spatial factors that introduce new challenges in understanding the causal effect of context. At the same time, activity space approaches allow for more accurate measurement of "treatment" effects (e.g., the "dose" of a given exposure). Moreover, the effects of exogenous changes in activity space environments (e.g., institutional changes) may be leveraged to assess contextual influences more precisely when specific information on geographic exposures is available.

It is important to acknowledge a number of limitations to our analyses. Our estimates of exposure to violent areas are based on nonhome time at the day level. Selection processes not accounted for in our models may be affecting the likelihood of experiencing nonhome time (including factors related to safety in the environment). Our analyses were also limited to both youth and routine activity locations within the City of Columbus. Although these locations account for the vast majority of nonhome time (90 percent), incorporating exposures beyond the City of Columbus would be necessary to estimate the true variability in exposure to violent areas. Including these noncity, mostly suburban exposures would likely also reveal more pronounced racial disparities.

An important next step in modeling exposure to violent areas would be to examine the amount of time youth spend in *highly* violent areas. The most violent areas of cities may be uniquely detrimental in their impact on youth and, considering exposure to the most dangerous neighborhoods, may reveal even more pronounced racial disparities.

Although activity space data introduce a range of new challenges, they also present new and exciting opportunities to researchers interested in the nature of environmental influence on youth development. Research on contextual effects will benefit tremendously from the ongoing introduction and refinement of new smartphone-based technologies for data collection and far more precise and extensive administrative and commercial data on the spatial environments to which urban youth are exposed.

Notes

1. Incorporation of controls for oversample areas result in minimal change to the results presented in Table 2.

2. This figure was determined empirically to capture the initiation of travel periods most accurately.

3. Although rape is typically reported in violent crime rates, the geocoded data available from Columbus Police do not identify rape separately from other types of sexual assault and sexual offences.

4. We employ eight years of data under the assumption that high crime rate areas establish reputations over a number of years. Consistently high crime areas are likely to engender the social processes (e.g., fear of crime and associated stress; expectations regarding informal social control) relevant for linking exposures to these areas with compromised health and development among youth. Averaging over a number of years also stabilizes violent crime rates at the (relatively small) block group level.

5. Although diversity and proportion African American are structurally dependent at extreme values, incorporating both does not result in multicollinearity in the analyses presented. Parameter estimates for the two variables change minimally when included separately or simultaneously.

6. We lose 198 days due to total time summing to greater than 1,440 minutes or lost timing data. These cases are due primarily to interviewer error in cleaning exit graphic timing data.

7. Roughly a fifth of eligible days were spent entirely at home. This is due, in part, to the fact that exit graphic days included three weekend days vs. only two weekdays. Dropping these days results in a potential selection problem if some families or youths systematically avoid exposure to nonhome locations, perhaps due to high levels of violence in the nonhome environment. To address this possibility, we estimated a logistic regression model of the likelihood of having one or more days entirely at home among youth, including a host of covariates. The predicted probability from this selection model was then included in models otherwise replicating those presented in Table 2. The estimates for the racial disparity in exposure to violent locations changed minimally across models.

8. As is typical in social surveys, missing data on household income are primarily responsible for the loss of cases due to listwise deletion.

9. Clustering of youth within block groups is limited. Employing census tracts as the neighborhood unit of analysis results in minimal change to the ICCs.

References

Anderson, Elijah. 1999. *Code of the street: Decency, violence, and the moral life of the inner city*. New York, NY: W. W. Norton.

Basta, Luke A., Therese S. Richmond, and Douglas J. Wiebe. 2010. Neighborhoods, daily activities, and measuring health risks experienced in urban environments. *Social Science & Medicine* 71 (11): 1943–50.

Bingenheimer, Jeffrey B., Robert T. Brennan, and Felton J. Earls. 2005. Firearm violence exposure and serious violent behavior. *Science* 308 (5726): 1323–26.

Bowen, Natasha K., and Gary L. Bowen. 1999. Effects of crime and violence in neighborhoods and schools on the school behavior and performance of adolescents. *Journal of Adolescent Research* 14 (3): 319–42.

Brenner, Allison B., Marc A. Zimmerman, Jose A. Bauermeister, and Cleopatra H. Caldwell. 2013. The physiological expression of living in disadvantaged neighborhoods for youth. *Journal of Youth and Adolescence* 42 (6): 792–806.

Browning, Christopher R., and Brian Soller. 2014. Moving beyond neighborhood: Activity spaces and ecological networks as contexts for youth development. *Cityscape* 16 (1): 165–96.

Browning, Christopher R., Brian Soller, Margo Gardner, and Jeanne Brooks-Gunn. 2013. "Feeling disorder" as a comparative and contingent process: Gender, neighborhood conditions, and adolescent mental health. *Journal of Health and Social Behavior* 54 (3): 296–314.

Buka, Stephen L., Theresa L. Stichick, Isolde Birdthistle, and Felton J. Earls. 2001. Youth exposure to violence: Prevalence, risks, and consequences. *American Journal of Orthopsychiatry* 71 (3): 298–310.

Carroll-Scott, Amy, Kathryn Gilstad-Hayden, Lisa Rosenthal, Susan M. Peters, Catherine McCaslin, Rebecca Joyce, and Jeannette R. Ickovics. 2013. Disentangling neighborhood contextual associations with child body mass index, diet, and physical activity: The role of built, socioeconomic, and social environments. *Social Science & Medicine* 95: 106–14.

Chan, Dara V., Christine A. Helfrich, Norman C. Hursh, E. Sally Rogers, and Sucharita Gopal. 2014. Measuring community integration using Geographic Information Systems (GIS) and participatory mapping for people who were once homeless. *Health & Place* 27: 92–101.

Chen, Edith, and Laurel Q. Paterson. 2006. Neighborhood, family, and subjective socioeconomic status: How do they relate to adolescent health? *Health Psychology* 25 (6): 704–14.

Clampet-Lundquist, Susan, and Douglas S. Massey. 2008. Neighborhood effects on economic self-sufficiency: A reconsideration of the moving to opportunity experiment. *American Journal of Sociology* 114 (1): 107–43.

Conger, Rand D., and M. Brent Donnellan. 2007. An interactionist perspective on the socioeconomic context of human development. *Annual Review of Psychology* 58 (1): 175–99.

Conger, Rand D., Xiaojia Ge, Glen H. Elder, Frederick O. Lorenz, and Ronald L. Simons. 1994. Economic stress, coercive family process, and developmental problems of adolescents. *Child Development* 65 (2): 541–61.

Cook, Thomas D., Melissa R. Herman, Meredith Phillips, and Richard A. Settersten, Jr. 2002. Some ways in which neighborhoods, nuclear families, friendship groups, and schools jointly affect changes in early adolescent development. *Child Development* 73 (4): 1283–309.

Elder, Glen H., Jr. 1998. The life course as developmental theory. *Child Development* 69 (1): 1–12.

Elder, Glen H., Jr., Monica Kirkpatrick Johnson, and Robert Crosnoe. 2003. The emergence and development of life course theory. In *Handbook of the life course*, eds. Jeylan T. Mortimer and Michael J. Shanahan, 3–19. New York, NY: Kluwer Academic/Plenum Publishers.

Fagan, Abigail A., Emily M. Wright, and Gillian M. Pinchevsky. 2015. Exposure to violence, substance use, and neighborhood context. *Social Science Research* 49: 314–26.

Felson, Marcus, and Lawrence E. Cohen. 1980. Human ecology and crime: A routine activity approach. *Human Ecology* 8 (4): 389–406.

Ford, Jodi L., and Christopher R. Browning. 2014. Effects of exposure to violence with a weapon during adolescence on adult hypertension. *Annals of Epidemiology* 24 (3): 193–98.

Foster, Holly, and Jeanne Brooks-Gunn. 2009. Toward a stress process model of children's exposure to physical family and community violence. *Clinical Child and Family Psychology Review* 12 (2): 71–94.

Fowler, Patrick J., Carolyn J. Tompsett, Jordan M. Braciszewski, Angela J. Jacques-Tiura, and Boris B. Baltes. 2009. Community violence: A meta-analysis on the effect of exposure and mental health outcomes of children and adolescents. *Development and Psychopathology* 21 (1): 227–59.

Graif, Corina, Andrew S. Gladfelter, and Stephen A. Matthews. 2014. Urban poverty and neighborhood effects on crime: Incorporating spatial and network perspectives. *Sociology Compass* 8 (9): 1140–55.

Hackman, Daniel A., Laura M. Betancourt, Nancy L. Brodsky, Hallam Hurt, and Martha J. Farah. 2012. Neighborhood disadvantage and adolescent stress reactivity. *Frontiers in Human Neuroscience* 6:277.

Hägerstrand, Torsten. 1970. What about people in regional science? *Papers of the Regional Science Association* 24 (1): 6–21.

Halfon, Neal, Kandyce Larson, Michael Lu, Ericka Tullis, and Shirley Russ. 2013. Lifecourse health development: Past, present and future. *Maternal and Child Health Journal* 18 (2): 344–65.

Jones, Malia, and Anne R. Pebley. 2014. Redefining neighborhoods using common destinations: Social characteristics of activity spaces and home census tracts compared. *Demography* 51 (3): 727–52.

Kirk, David S. 2009. Unraveling the contextual effects on student suspension and juvenile arrest: The independent and interdependent influences of school, neighborhood, and family social controls. *Criminology* 47 (2): 479–520.

Kirk, David S., and Margaret Hardy. 2014. The acute and enduring consequences of exposure to violence on youth mental health and aggression. *Justice Quarterly* 31 (3): 539–67.

Kubrin, Charis E., and Ronald Weitzer. 2003. New directions in social disorganization theory. *Journal of Research in Crime and Delinquency* 40 (4): 374–402.

Kupersmidt, Janis B., Ariana Shahinfar, and Mary Ellen Voegler-Lee. 2002. Children's exposure to community violence. In *Helping children cope with disasters and terrorism*, eds. Anette M. La Greca, Wendy K. Silverman, Eric M. Vernberg, and Michael C. Roberts, 381–401. Washington, DC: American Psychological Association.

Kwan, Mei-Po. 2009. From place-based to people-based exposure measures. *Social Science & Medicine* 69 (9): 1311–13.

Kwan, Mei-Po. 2013. Beyond space (as we knew it): Toward temporally integrated geographies of segregation, health, and accessibility. *Annals of the Association of American Geographers* 103 (5): 1078–86.

Margolin, Gayla, and Elana B. Gordis. 2000. The effects of family and community violence on children. *Annual Review of Psychology* 51 (1): 445–79.

Margolin, Gayla, and Elana B. Gordis. 2004. Children's exposure to violence in the family and community. *Current Directions in Psychological Science* 13 (4): 152–55.

Massey, Douglas S. 2004. Segregation and stratification: A biosocial perspective. *Du Bois Review: Social Science Research on Race* 1 (1): 7–25.

Massey, Douglas S., and Stefanie Brodmann. 2014. *Spheres of influence: The social ecology of racial and class inequality*. New York, NY: Russell Sage Foundation.

Mayer, Susan E. 1997. *What money can't buy: Family income and children's life chances*. Cambridge, MA: Harvard University Press.

McEwen, Bruce S., and Eliot Stellar. 1993. Stress and the individual: Mechanisms leading to disease. *Archives of Internal Medicine* 153 (18): 2093–101.

McKenzie, R. D. 1921. The neighborhood: A study of local life in the city of Columbus, Ohio. II. *American Journal of Sociology* 27 (3): 344–63.

McMahon, Susan D., Nathan R. Todd, Andrew Martinez, Crystal Coker, Ching-Fan Sheu, Jason Washburn, and Seema Shah. 2013. Aggressive and prosocial behavior: Community violence, cognitive, and behavioral predictors among urban African American youth. *American Journal of Community Psychology* 51 (3–4): 407–21.

Moos, R., and B. Moos. 1994. *Family environment scale manual: development, applications, research*. 3rd ed. Palo Alto, CA: Consulting Psychologist Press.

Osofsky, Howard J., and Joy D. Osofsky. 2004. Children's exposure to community violence: Psychoanalytic perspectives on evaluation and treatment. In *Analysts in the trenches: Streets, schools, war zones*, eds. Bruce Sklarew, Stuart W. Twemlow, and Sallye M. Wilkinson, 237–256. Burlingame, CA: Analytic Press.

Osofsky, Joy D. 1999. The impact of violence on children. *The Future of Children* 9 (3): 33–49.

Owens, Ann. 2010. Neighborhoods and schools as competing and reinforcing contexts for educational attainment. *Sociology of Education* 83 (4): 287–311.

Pattillo-McCoy, Mary. 1999. *Black picket fences: Privilege and peril among the black middle class*. Chicago, IL: University of Chicago Press.

Pearlin, Leonard I., Scott Schieman, Elena M. Fazio, and Stephen C. Meersman. 2005. Stress, health, and the life course: Some conceptual perspectives. *Journal of Health and Social Behavior* 46 (2): 205–19.

Rudolph, Kara E., Gary S. Wand, Elizabeth A. Stuart, Thomas A. Glass, Andrea H. Marques, Roman Duncko, and Kathleen R. Merikangas. 2014. The association between cortisol and neighborhood disadvantage in a U.S. population-based sample of adolescents. *Health & Place* 25:68–77.

Sampson, Robert J. 2012. *Great American city*. Chicago, IL: University of Chicago Press.

Sampson, Robert J., Jeffrey D. Morenoff, and Thomas Gannon-Rowley. 2002. Assessing "neighborhood effects": Social processes and new directions in research. *Annual Review of Sociology* 28:443–78.

Sampson, Robert J., Jeffrey D. Morenoff, and Stephen W. Raudenbush. 2005. Social anatomy of racial and ethnic disparities in violence. *American Journal of Public Health* 95 (2): 224–32.

Sampson, Robert J., Stephen W. Raudenbush, and Felton J. Earls. 1997. Neighborhoods and violent crime: A multilevel study of collective efficacy. *Science* 277 (5328): 918–24.

Selner-O'Hagan, Mary Beth, Daniel J. Kindlon, Stephen L. Buka, Stephen W. Raudenbush, and Felton J. Earls. 1998. Assessing exposure to violence in urban youth. *Journal of Child Psychology and Psychiatry* 39 (2): 215–24.

Sharkey, Patrick. 2010. The acute effect of local homicides on children's cognitive performance. *Proceedings of the National Academy of Sciences* 107 (26): 11733–38.

Sharkey, Patrick T., Nicole Tirado-Strayer, Andrew V. Papachristos, and C. Cybele Raver. 2012. The effect of local violence on children's attention and impulse control. *American Journal of Public Health* 102 (12): 2287–93.

Sharp, Gregory, Justin T. Denney, and Rachel T. Kimbro. 2015. Multiple contexts of exposure: Activity spaces, residential neighborhoods, and self-rated health. *Social Science & Medicine* 146:204–13.

Shonkoff, Jack P., and Andrew S. Garner, Committee on Psychosocial Aspects of Child and Family Health, Committee on Early Childhood, Adoption, and Dependent Care, and Section on Developmental and Behavioral Pediatrics. 2012. The lifelong effects of early childhood adversity and toxic stress. *Pediatrics* 129 (1): e232–e246.

Simons, Leslie Gordon, K. A. S. Wickrama, T. K. Lee, Melissa Landers-Potts, Carolyn Cutrona, and Rand D. Conger. 2016. Testing family stress and family investment explanations for conduct problems among African American adolescents. *Journal of Marriage and Family* 78 (2): 498–515.

Skogan, Wesley G. 1990. *Disorder and decline: Crime and the spiral of decay in American neighborhoods*. New York, NY: Free Press.

Small, Mario L. 2009. *Unanticipated gains: Origins of network inequality in everyday life*. Oxford: Oxford University Press.

Sternthal, Michelle, H. J. Jun, Felton J. Earls, and Rosalind J. Wright. 2010. Community violence and urban childhood asthma: a multilevel analysis. *European Respiratory Journal* 36 (6): 1400–409.

Teitler, Julien O., and Christopher C. Weiss. 2000. Effects of neighborhood and school environments on transitions to first sexual intercourse. *Sociology of Education* 73 (2): 112–32.

Theall, Katherine P., Zoë H. Brett, Elizabeth A. Shirtcliff, Erin C. Dunn, and Stacy S. Drury. 2013. Neighborhood disorder and telomeres: Connecting children's exposure to community level stress and cellular response. *Social Science & Medicine* 85: 50–58.

Turner, R. Jay, B. Gail Frankel, and Deborah M. Levin. 1983. Social support: Conceptualization, measurement, and implications for mental health. In *Research in Community & Mental Health*, vol. 3, ed. James R. Greenley, 67–111. Greenwich, CT: JAI Press.

Vanfossen, Beth, C. Hendricks Brown, Sheppard Kellam, Natalie Sokoloff, and Susan Doering. 2010. Neighborhood context and the development of aggression in boys and girls. *Journal of community psychology* 38 (3): 329–49.

Vliegenthart, J., G. Noppe, E. F. C. van Rossum, J. W. Koper, H. Raat, and E. L. T. van den Akker. 2016. Socioeconomic status in children is associated with hair cortisol levels as a biological measure of chronic stress. *Psychoneuroendocrinology* 65:9–14.

Wickrama, K. A. S., Samuel Noh, and Chalandra M. Bryant. 2005. Racial differences in adolescent distress: Differential effects of family and community for blacks and whites. *Journal of Community Psychology* 33:261–82.

Wiebe, Douglas J., Therese S. Richmond, Wensheng Guo, Paul D. Allison, Judd E. Hollander, Michael L. Nance, and Charles C. Branas. 2016. Mapping activity patterns to quantify risk of violent assault in urban environments. *Epidemiology* 27 (1): 32–41.

Wikström, Per-Olof H., Dietrich Oberwittler, Kyle Treiber, and Beth Hardie. 2012. *Breaking rules: The social and situational dynamics of young people's urban crime*. Oxford: Oxford University Press.

Wilson, William J. 1987. *The truly disadvantaged: The inner city, the underclass, and public policy*. Chicago, IL: University of Chicago Press.

Yin, Li, Samina Raja, Xiao Li, Yuan Lai, Leonard Epstein, and James Roemmich. 2013. Neighbourhood for playing: Using GPS, GIS and accelerometry to delineate areas within which youth are physically active. *Urban Studies* 50 (14): 2922–39.

Zimmerman, Gregory M., and Steven Messner. 2013. Individual, family background, and contextual explanations of racial and ethnic disparities in youths' exposure to violence. *American Journal of Public Health* 103:435–42.

Linking Federal Surveys with Administrative Data to Improve Research on Families

By
AMY O'HARA,
RACHEL M. SHATTUCK,
and
ROBERT M. GOERGE

Linkage of federal, state, and local administrative records to survey data holds great promise for research on families, in particular research on low-income families. Researchers can use administrative records in conjunction with survey data to better measure family relationships and to capture the experiences of individuals and family members across multiple points in time and social and economic domains. Administrative data can be used to evaluate program participation in government social welfare programs, as well as to evaluate the accuracy of reporting on receipt of such benefits. Administrative records can also be used to enhance collection and accuracy of survey and census data and to improve coverage of hard-to-reach populations. This article discusses potential uses of linked administrative and survey data, gives an overview of the linking methodology and infrastructure (including limitations), and reviews social science literature that has used this method to date.

Keywords: administrative records; combining multiple data sources; family research; methodology

Family is a central and enduring social institution. Whether in the form of a nuclear family, extended family, childless couple, or one of many other household structures, families are fundamental building blocks of neighborhoods, religious communities, and social networks. Because

Amy O'Hara is chief of the Center for Administrative Records Research and Applications at the U.S. Census Bureau. She addresses legal, policy, and methodological issues to expand use of administrative records data in federal statistics.

Rachel M. Shattuck is a sociologist and demographer in the Center for Administrative Records Research and Applications at the U.S. Census Bureau. Among her current projects is an analysis of childcare records linked to subsequent educational outcomes in survey data.

Robert M. Goerge is a senior research fellow at Chapin Hall at the University of Chicago and has senior fellow positions at the University of Chicago Harris School of Public Policy and Computation Institute and NORC.

Correspondence: rachel.m.shattuck@census.gov

DOI: 10.1177/0002716216678391

of the pivotal role that family plays in child development and socialization and its intersections with domains such as gender, socioeconomic status, race and ethnicity, paid work, care work, living arrangements, and the law, it is a topic of interest to social scientists and policy-makers alike (Gornick and Meyers 2003). As families and family members pass through the life course, they interact with multiple institutions, generating administrative data on their encounters and outcomes. Linking such administrative data with existing research data, including survey and census data, provides great opportunity to understand families.

The U.S. Census Bureau accesses data from federal agencies, including the Internal Revenue Service (IRS), Department of Housing and Urban Development (HUD), Social Security Administration (SSA), and Centers for Medicare and Medicaid Services, as well as state-level data on programs such as the Supplemental Nutrition Assistance Program (SNAP), the Special Supplemental Nutrition Program for Women, Infants, and Children (WIC), and the Temporary Assistance for Needy Families (TANF) program, among others. The Census Bureau links these files at the person level with data from household surveys such as the Survey of Income and Program Participation (SIPP), the Current Population Survey (CPS), and the American Community Survey (ACS) as well as with decennial census data. We discuss this linking process below.

Linking administrative data to survey data holds a great deal of promise for family researchers. For example, administrative data can help researchers to understand individual family members' labor force experiences (IRS data), health insurance coverage (Medicare and Medicaid data), and interactions with the criminal justice system (corrections data), as well as outcomes for veterans and military families (Veterans Affairs data). When linked to survey data, these records may allow researchers to observe how the experiences of one family member affect other members of the family.

Administrative data are particularly valuable for understanding low-income families, who are often underrepresented in survey data. For example, records generated by programs such as TANF, SNAP, and WIC can help researchers to understand how low-income families meet their material needs and how well these programs serve their intended populations. Administrative data can also help to illuminate family relationships. With increasing family complexity, outcomes for individuals and families may need to be observed across multiple households (Manning, Brown, and Stykes 2014) and to include members whose relationships to one another are not easily captured by surveys. Such complex relationships may sometimes be better understood by combining information from administrative and survey data. Individual family members experience change across their own lives, and family composition changes over time as unions are formed and dissolved, children are born, and family members die. Administrative data—which may capture information about an individual at multiple points in time—can help to extend the chronological reach of survey data.

In this article, we discuss the potential contributions to family research of linking administrative and survey data. We briefly review some of the economic, sociological, and policy research that to date has productively used linked administrative data and survey data to address five research questions. We discuss the process of record linkage and describe the administrative infrastructure available at the U.S. Census Bureau to facilitate researchers' access to these data. Finally, we discuss some data limitations endemic to administrative data, and the potential for future research in this area.

Framework

The value of linking administrative data to survey data to study family issues is best demonstrated by citing some of the research that has been conducted using this method. The increased prevalence of multipartner fertility, complex families, multigenerational households, and household doubling up challenges researchers to capture and understand family and household relationships. Survey and census data may capture only one dimension of an interrelationship and may not capture all members of a family that spans multiple households.

This review of a selection of literature investigates whether and how the linkage of administrative records and survey data can address five questions highlighting family research challenges.

1. Can administrative data generate new ways of measuring individual families and/or households?
2. Can administrative data be used to extend information on families in survey and decennial census data?
3. Can administrative data be used to evaluate accuracy of survey data?
4. Can administrative data be used to evaluate coverage in census records of hard-to-reach populations who may be better represented in administrative data?
5. Can administrative data be used to evaluate access to and participation in social welfare programs?

By answering these questions, we highlight how linkages between administrative records and survey data can create broader and deeper data resources. In the section that follows, we describe data sources and linkage methods that permit researchers to undertake analyses that address these five questions.

Data and Methods

The U.S. Census Bureau links administrative data at the person and address level to information collected by the Bureau, including data from SIPP, the CPS, and the ACS, as well as decennial census data. Using common fields across sources, probabilistic matching results in a unique identifying number called a protected identification key and strips records of identifying details. Unique identifying numbers called master address file identifiers are also appended to facilitate linkages by address. These persistent unique identifiers enable data records for the same person to be linked across files and over time while preserving confidentiality. The Census Bureau's authority to access records with personal identifiers and linkage infrastructure enable researchers to identify family relationships in multiple data sources and follow family members and their experiences in multiple domains. The Census Bureau is also involved in a joint effort with the National Academy of Sciences and academic researchers to digitize and match names from the 1990 Decennial Census to administrative data. This pilot project is part of the American Opportunity Study, seeking linkages of census, survey, and administrative data across generations to study social and economic mobility (Grusky, Smeeding, and Snipp 2015).

Chapin Hall at the University of Chicago maintains an integrated database on child and family programs for the state of Illinois. Chapin Hall's database spans education, employment, income support, and health domains, including programs such as public school, pre-K, and Head Start data; unemployment insurance wage and benefit records; Workforce Innovation and Opportunity Act data; Supplemental Security Income and Aid to the Aged, Blind, or Disabled; TANF and TANF work programs; SNAP; childcare subsidy and licensing data; and data on Medicaid eligible individuals, Medicaid providers, and Medicaid claims. This collection of rich data has been carefully curated for decades and allows for deep and diverse studies in Illinois.

Although the Census Bureau currently has a large collection of national administrative data files, the Bureau is also seeking to expand its collection of state-level human services program data from programs such as SNAP, WIC, and TANF. The Bureau hopes to eventually acquire data from all states to provide additional, state-level information for researchers, in particular on low-income families and individuals.

Discussion

Linking administrative data with survey data can contribute to expanding and improving family research. Below we discuss papers that illustrate how administrative data can most productively be used to study families, in ways that address our five research questions. Table 1 recaps our research questions and summarizes examples of data sources that can be used to generate answers to these questions.

Improving measurement of families and households

The measurement of families has evolved over time in both censuses and sample surveys (Ruggles and Brower 2003). To measure families in sample surveys, the Census Bureau currently captures information about householders and other persons in a given housing unit. The amount of detail that associates a household's residents with the householder varies across surveys. Some surveys contain rich relationship pointers (CPS, SIPP), allowing analysts to determine relationships for families and subfamilies (Kennedy and Fitch 2012). Other surveys, such as the ACS, only measure relationships of household residents to the reference person.

TABLE 1
Synopsis of Research Questions and Data Sources

Research question	Examples of federal survey/census and administrative data used by the U.S. Census Bureau
Can administrative data generate new ways of measuring individual families and/or households?	Decennial census, household surveys, HUD, IRS, SNAP, TANF
Can administrative data be used to extend information on families in survey and decennial census data?	Decennial census, household surveys, HUD, IRS, Medicaid, SNAP, SSA, WIC
Can administrative data be used to evaluate accuracy of survey data?	Household surveys, HUD, IRS, Medicaid, Medicare, TANF, SNAP, SSA, WIC
Can administrative data be used to evaluate coverage in census records of hard-to-reach populations who may be better represented in administrative data?	Decennial census, household surveys, Indian Health Service, IRS, Medicaid, Medicare, SNAP, TANF, WIC
Can administrative data be used to evaluate access to and participation in social welfare programs?	Household surveys, SNAP, TANF, WIC

Previous research by Heggeness, Alexander, and Stern (2012) has demonstrated that differences in how family units are measured can affect larger estimates, such as household incidence of poverty. Administrative records allow for ways of measuring family units that complement family measurement in survey files, particularly with respect to measurement of household-level resources. Some administrative data sources are person level, such as information returns from the IRS. For example, wage statements (Form W-2) reflect earner information. In contrast, individual income tax returns (Form 1040) include tax filing units, which may include a secondary filer and tax dependents. This forms a "family" unit of sorts, using conditions on coresidence and support defined by the Internal Revenue Code. Similarly, data from SNAP and TANF programs include household units, reported to the human services agencies by program applicants. SNAP units are defined as groups of people who live together and purchase and prepare meals together (Harris 2014). TANF rules vary from state to state as to whether pregnant women are eligible, and whether teen parents are considered head of their own family unit or must be counted with their own parents (Huber et al. 2015). These alternative ways of measuring family relationships can augment the understanding of family units as defined in survey data.

Extending family information from survey and census records

Families present significant measurement challenges for social scientists. There are many types of families (e.g., two-parent families, single-parent families, no-children families, same-sex parents, grandparent-headed families, and so

FIGURE 1
Family Relationships as Reflected in Survey and Administrative Data

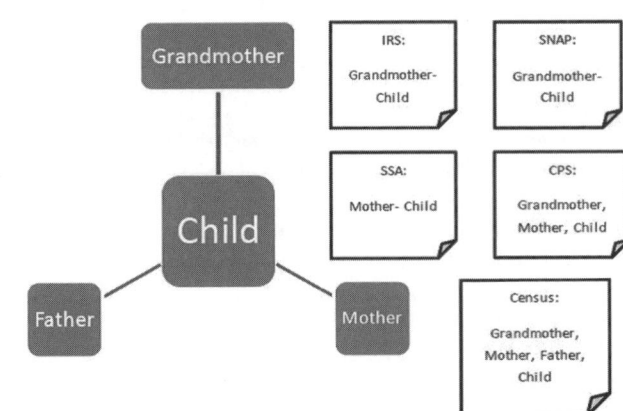

on). Membership of families changes as new members are born or adopted and as others separate. Family members may all live in the same place (or not) for different periods of time (Manning, Brown, and Stykes 2014; Wall and Bolzman 2013). Multiple subfamilies may live in the same household, and also compose extended or multigenerational families (Mykyta and McCartney 2011). Finally, not all families are related by blood, as children may be formally or informally adopted, or individuals may identify as fictive kin (Radel, Bramlett, and Waters 2010). Families may have complex structures that vary over time and context. Thus, household membership may or may not align with family membership.

Different household members may be observed in different survey and administrative data. Figure 1 shows a stylized example of a family centered on a child. Assume the child lives in the maternal grandmother's house. The child's mother is often present, and the child's father is occasionally present. The node between the child and maternal grandmother is observed in census and survey contexts where questions are asked of the householder. The other nodes connecting the child with his or her mother and father do not appear in the survey. The node linking the mother and child appears in SSA and state vital records data. The Social Security and vital records may also include the node between the child and his or her father. In other sources that may provide more information about relationships—such as public housing or voucher data from HUD, Medicaid, food security support from SNAP or WIC, or IRS Earned Income Tax Credit EITC—the family relationships reflected will depend on whether the grandmother or mother was the responsible party who enrolled the child.

Census Bureau researchers have been generating research on how individuals, especially children, are captured in administrative data, and how family structures differ from self-reported information collected in survey or population census data. These comparisons can be used to improve survey accuracy and to offer new information about respondents at various points in their lives. Research matching decennial census data to administrative data uses probabilistic linkage infrastructure at the Census Bureau to identify individuals in each data source that are likely to be the same person and/or people. Luque and Wagner (2015) compare parent-child links constructed from SSA data using probabilistic name matching to filer-dependent child data from individuals' IRS income tax returns and parent-child relationships from the 2010 U.S. Census. Parent-child information from IRS data can be used to supplement information from census data. Luque and Wagner (2015) argue that more accurately linking parents to children in administrative and survey data can improve researchers' ability to study intergenerational mobility, labor market outcomes, and child development. Massey (2014) shows how record linkage can be used to identify family relationships over time. She finds that data from the 1960 U.S. Census can be linked to current administrative data to study individuals' outcomes over the life course as well as their relationships to other family members. Using a transcribed sample of the 1960 U.S. Census with very limited identifiers (name, quarter of birth, and year of birth), the inclusion of parent name information increases the match rate to the administrative data.

Evaluating accuracy of survey data

By linking administrative data on participants in social programs to survey data, researchers can evaluate the accuracy of reporting on participation in those programs. Recipients of public benefits such as TANF, SNAP, WIC, Medicaid, and Medicare commonly underreport their receipt of these benefits, either because of confusion about the nature of the benefit that they are receiving (Marquis and Moore 1990) or because of desirability bias (Nederhof 1985). Meyer and Mittag (2015) link the CPS with New York State SNAP, TANF, and housing subsidy data to examine the coverage of the safety net, specifically the share of people without work or program receipt. They find that survey data both understate the income of low-income households and overestimate the share of single mothers who are neither working nor receiving benefits. Similarly, Meyer and Goerge (2011) find underreporting of food stamp benefits by roughly 35 percent in the ACS and 50 percent in the SIPP. Card et al. (2004) use California state-level Medicare data to show that the SIPP underreports Medicare receipt by about 10 percent, with the highest underreporting about young children. Huynh, Rupp, and Sears (2002) match national-level SSA data to the SIPP to show that the SIPP underestimates the dollar amount of benefits received. Kim and Tamborini (2014) match SIPP and IRS W-2 data to show that SIPP respondents misreport their earnings, with different patterns of misreporting among different educational and race/ethnic groups. As the U.S. Census Bureau and Chapin Hall acquire more administrative data sources and make them available to researchers, we expect that researchers will be able to further examine survey-administrative record discrepancies, thereby improving the understanding of the strengths and limitations of survey data for all researchers.

Evaluating coverage of hard-to-reach individuals and families

The availability of high-quality data for social science research has long relied on federally funded large-scale longitudinal surveys in education, health, and human services. Surveys sponsored by the U.S. Census Bureau, Department of Health and Human Services, Department of Agriculture, and other federal agencies support traditional research data in the social sciences. These surveys face declining respondent cooperation and budget constraints, and sponsors are turning to auxiliary and complementary sources of information. Administrative data can be used to improve imputation as a way around unit and item nonresponse (Meyer, Mok, and Sullivan 2015) and to evaluate the accuracy of existing data. The Census Bureau is exploring methods to use administrative data instead of asking respondents particular questions, especially those that are considered intrusive or are difficult to answer. For example, the Census Bureau is assessing the extent to which IRS income records could replace the ACS income questions. Data quality, coverage, and timing issues must be analyzed, but in the future, IRS data may make data collection for some subpopulations unnecessary.

Substantial internal research at the Census Bureau has been dedicated to evaluating administrative data coverage of various subpopulations in decennial census records. Some groups of people—such as children (O'Hare 2015), men, and African Americans (Clogg, Massagli, and Eliason 1989; Raley 2002)—are known to be systematically missed and undercounted by censuses and sample surveys. Internal studies have evaluated how well individuals from national-level Indian Health Service, IRS, Medicaid, and Medicare files—and state-level SNAP and WIC files—are represented in 2010 U.S. Census files, and how much information matches across these files. This coverage and quality information can improve survey sampling frames, reduce cost and burden during the 2020 Decennial Census, and may also be used to supplement incomplete records and improve imputation in the case of unit and item nonresponse. In addition, research has evaluated whether hard-to-reach and frequently underreported groups may be better represented in administrative data than in decennial census records. Results from internal research show, for example, that supplementing census records with WIC and SNAP records substantially improves coverage of young children.

Evaluating social welfare program access and participation

Linking survey and administrative data allows comparison of household and benefit units. This allows researchers to evaluate the extent to which individuals who and households that are estimated to be eligible for benefits actually take advantage of them. Such information can help program administrators to better target their services to eligible populations and nonparticipating individuals by generating information about which characteristics are associated with increased likelihood of nonparticipation. In the future, better understanding of household- and family-level patterns of benefit usage may help survey methodologists to design questionnaires to prompt respondents' memories in ways that will improve data accuracy. Czajka, Cunnyngham, and Rosso (2015) link SNAP administrative data from New York and Colorado to three surveys, the ACS, CPS, and SIPP. They compare SNAP unit membership as recorded in the administrative data to simulated unit membership (ACS and CPS) and reported unit membership (SIPP). They show that in 50 percent of survey households, all members of both the ACS survey household and the New York administrative unit were matched to the other dataset, and that the number of simulated and administrative SNAP units were the same. The vast majority of these households contained only one SNAP unit. In addition, between 44 and 47 percent of the households receiving SNAP benefits also included SNAP nonparticipants.

Also using the ACS, Scherpf, Newman, and Prell (2015) link survey data to state SNAP records to study how well SNAP is reaching the intended population. They find that substituting SNAP receipt as measured in administrative data for survey reports of SNAP receipt increases the proportion of lower income units receiving SNAP benefits in a year. More individuals and families receive SNAP than survey research would indicate. Looking at total benefit amounts and number of months of receipt in the year reveals that lower income units use SNAP more intensively. Newman and Scherpf (2013) use ACS and state-level data to examine uptake of SNAP in Texas, according to different geographic areas and demographic groups. They find that eligible households with children had higher access rates than other groups, and that eligible people aged 65 and up had the lowest access rates.

Limitations

Our approach to improving family research by linking federal surveys and administrative data relies not only on access to federal and state program files but also on the ability to conduct analyses and publish findings while maintaining the privacy of individuals. This section describes these access and confidentiality challenges—as well as other data access barriers facing researchers—and the inherent limitations of using administrative records for research purposes.

Both Chapin Hall and the U.S. Census Bureau have been acquiring and integrating administrative data for decades. This involves the engagement and cooperation of the local, state, or federal agency that owns the data. The data are often trapped in silos that span levels of government (e.g., federal, state, county, and city), silos that exist within and across agencies, and silos across domains (e.g., welfare and benefit programs, human services, law enforcement, education, employment/wages, and public health programs). Data access cannot occur without agreement from lawyers on permissions for data use and the cooperation of program and information technology staff to define, extract, and transmit data. Once studies are prepared, researchers must also engage program administrators and sometimes obtain their permission to release the findings.

At the Census Bureau, all record linkages must support its programs by improving census and survey collection and accuracy, or supporting other statistical activity. When the Census Bureau is deciding whether to link datasets, the linkage's potential to support the agency's mission is evaluated along with other criteria including sensitivity, cost, burden, timeliness, and data quality. Protecting the confidentiality of individuals described in these data is critical; at the Census Bureau, all personal and business information acquired by the Census Bureau and any resulting linkages are protected by law under Title 13 of the U.S. Code.

Once the datasets are linked, academic researchers from across disciplines, such as economics, policy analysis, and sociology, may want access to the linked data. Federal, state, and local government program administrators who wish to evaluate their own program implementation or policy impacts may also want access to the data. Obtaining permission to use the data and gaining access to the data are difficult and time-consuming for researchers. To help ameliorate these challenges, the Census Bureau brings administrative databases from different institutions and domains together in one place. By leveraging existing systems for governance and privacy protection, the Census Bureau is expanding secure access to data for researchers through its existing network of Federal Statistical Research Data Centers. The Census Bureau is working to improve the infrastructure by which researchers gain access to data. For example, the Bureau is creating user interfaces that allow for the use of comprehensive metadata, as well as robust protocols that allow for access to data integrated across surveys, to federal administrative data, and to state and local administrative data. These efforts align with the charge of the Evidence-Based Policymaking Commission, which was established under Public Law 114-140 to study how administrative data on federal programs, survey data, and other statistical data series can inform program evaluation, cost-benefit analyses, and policy-relevant studies.

Finally, while linked administrative and survey data have unique characteristics and benefits for researchers, they also have unique shortcomings. For example, a major advantage of administrative records is that they include information on the full population receiving a particular benefit or service at a given time, and are not subject to the uncertainty inherent in survey reporting. However, they are representative of neither the entire U.S. population nor even the entire population that may be eligible for a benefit or service. Another major advantage is that administrative data may include different ways of measuring family units and may contain information about families that can augment the information available in survey research. However, administrative data include many fewer variables than typically appear in survey data, so the extent to which they can augment survey data may be inconsistent. Because administrative data have not been designed explicitly for use in social science research, they may contain systematic or random error.

Conclusion

The dynamic nature of family and the expanding number of potential social research and policy questions concerning family call for new and different research approaches to studying families. Linking administrative data to existing survey data is one methodological innovation that offers researchers the capacity to enhance, extend, and improve survey data.

Given the growing electronic data collection around many facets of individuals' lives, researchers can make use of administrative data to study not only individuals in families, but also the family unit itself, whatever form it might take. By pursuing administrative data sources, making these files usable for researchers, and providing a platform by which internal and external researchers may use these data, the Census Bureau and research partners, such as Chapin Hall, hope to launch a new era in quantitative research on families. As discussed above, linkage of administrative data with survey data allows for better measurement of family units and different ways to understand interrelationships in families and households (for example, in terms of food and meal production, social supports, and financial resources). This will help researchers to better meet the measurement challenges around family identification and to understand family systems and the functioning of families as entities within local, global, social, and economic contexts. Linking administrative records to survey data may also allow researchers to observe individuals and their needs at multiple points in time, thus tracing their outcomes over longer periods than are typically available outside of longitudinal studies. Linking administrative data and survey data allows researchers to trace family outcomes over time as well—for example, observing links between children, parents, and grandparents, their use of social services, and their near- and long-term social and economic mobility. We believe that the availability of these data will spark the creation of new and innovative methods; generate new knowledge to the benefit of social science research; and improve federal, state, and local policymaking.

References

Card, David, Andrew K. G. Hildreth, and Lara D. Shore-Sheppard. 2004. The measurement of Medicaid coverage in the SIPP: Evidence from matched records. *Journal of Business and Economic Statistics* 22:410–20.

Clogg, Clifford C., Michael P. Massagli, and Scott R. Eliason. 1989. Population undercount and social science research. *Social Indicators Research* 21:559–98.

Czajka, John L., Karen Cunnyngham, and Randy Rosso. 2015. Simulated versus actual SNAP unit composition in survey households in two states. Paper presented at the 2015 Federal Committee on Statistical Methodology Research Conference. Available from http://fcsm.sites.usa.gov/files/2016/03/I1_Cunnyngham_2015FCSM.pdf.

Gornick, Janet C., and Marcia K. Meyers. 2003. *Families that work: Policies for reconciling parenthood and employment*. New York, NY: Russell Sage Foundation.

Grusky, David B., Timothy M. Smeeding, and C. Matthew Snipp. 2015. A new infrastructure for monitoring social mobility in the United States. *The ANNALS of the American Academy of Political and Social Science* 657:63–82.

Harris, Benjamin Cerf. 2014. *SNAP eligibility and take-up among veteran households during the Great Recession: Evidence from New York SNAP administrative records linked to the ACS*. Washington, DC: Center for Administrative Records Research and Applications, U.S. Census Bureau.

Heggeness, Misty L., Trent Alexander, and Sharon Stern. 2012. Alternative strategies for grouping people into resource units: Measuring poverty in the American Community Survey. SEHSD Working Paper Number WP-2012-09. Washington, DC: U.S. Census Bureau.

Huber, Erika, Elissa Cohen, Amanda Briggs, and David Kasabian. 2015. *Welfare rules databook: State TANF policies as of July 2014*. Washington, DC: The Urban Institute.

Huynh, Minh, Kalman Rupp, and James Sears. 2002. The assessment of Survey of Income and Program Participation (SIPP) benefit data using longitudinal administrative records. Survey of Income and Program Participation Working Paper #238. Washington, DC: U.S. Census Bureau.

Kennedy, Sheela, and Catherine A. Fitch. 2012. Measuring family structure and cohabitation in the United States: Assessing the impact of new data from the Current Population Survey. *Demography* 49:1479–98.

Kim, ChangHwan, and Christopher R. Tamborini. 2014. Response error in earnings: An analysis of the Survey of Income and Program Participation matched with administrative data. *Sociological Methods and Research* 43:39–72.

Luque, Adela, and Deborah Wagner. 2015. Assessing coverage and quality of the 2007 Prototype Census Kidlink Database. Center for Administrative Records Research and Applications Working Paper Series #2015-07. Washington, DC: U.S. Census Bureau.

Manning, Wendy D., Susan L. Brown, and J. Bart Stykes. 2014. Family complexity among children in the United States. *The ANNALS of the American Academy of Political and Social Science* 654:48–65.

Marquis, Kent H., and Jeffrey C. Moore. 1990. Measurement errors in SIPP program reports. Proceedings of the 1990 Annual Research Conference, 721–45. Washington, DC: U.S. Census Bureau.

Massey, Catherine G. 2014. Creating linked historical data: An assessment of the Census Bureau's ability to assign protected identification keys to the 1960 Census. Center for Administrative Records Research and Applications Working Paper #2014-12. Washington, DC: U.S. Census Bureau.

Meyer, Bruce D., and Robert M. Goerge. 2011. Errors in survey reporting and imputation and their effects on estimates of food stamp program participation. Center for Economic Studies Paper #CES-WP-11-14. Washington, DC: U.S. Census Bureau.

Meyer, Bruce D., and Nikolas Mittag. 2015. Using linked survey and administrative data to better measure income: Implications for poverty, program effectiveness and holes in the safety net. National Bureau of Economic Research Working Paper #21676. Cambridge, MA.

Meyer, Bruce D., Wallace K. C. Mok, and James X. Sullivan. 2015. Household surveys in crisis. National Bureau of Economic Research Working Paper #21399. Cambridge, MA.

Mykyta, Larissa, and Suzanne McCartney. 2011. The effects of recession on household composition: "Doubling up" and economic well-being. SEHSD Working Paper Number 2011-4. Washington, DC: U.S. Census Bureau.

Nederhof, Anton J. 1985. Methods of coping with social desirability bias: A review. *European Journal of Social Psychology* 15:263–80.

Newman, Constance, and Erik Scherpf. 2013. Supplemental Nutrition Assistance Program (SNAP) access at state and county levels. Economic Research Report #156. Washington, DC: U.S. Department of Agriculture Economic Research Service.

O'Hare, William P. 2015. Coverage of young children in the 2010 decennial census. In *The undercount of young children in the U.S. Decennial Census*, 27–37. Heidelberg: Springer.

Radel, Laura F., Matthew D. Bramlett, and Annette Waters. 2010. Legal and informal adoption by relatives in the U.S.: Comparative characteristics and well-being from a nationally- representative sample. *Adoption Quarterly* 13:268–91.

Raley, R. Kelly. 2002. The effects of differential undercount on survey estimates of race differences in marriage. *Journal of Marriage and Family* 64:774–79.

Ruggles, Steven, and Susan Brower. 2003. Measurement of household and family composition in the United States, 1850–2000. *Population and Development Review* 29:73–101.

Scherpf, Erik, Constance Newman, and Mark Prell. 2015. Improving the assessment of SNAP targeting using administrative records. Economic Research Report #186. Washington, DC: U.S. Department of Agriculture Economic Research Service.

Wall, Karin, and Claudio Bolzman. 2013. Mapping the new plurality of transnational families: A life course perspective. In *Transnational families, migration and the circulation of care work*, Loretta Baldasar and Laura Merla, eds. New York, NY: Routledge.

Predicting Asthma Prevalence by Linking Social Media Data and Traditional Surveys

By
HONGYING DAI,
BRIAN R. LEE,
and
JIANQIANG HAO

Asthma is one of the most common chronic diseases that has a profound impact on people's well-being and our society. In this study, we link multiple large-scale data sources to construct an epidemiological model to predict asthma prevalence across geographic regions. We use: (1) the Social Media Monitoring (SMM) data from Twitter (N = 500 million tweets/day), (2) the 2014 Behavioral Risk Factor Surveillance System (BRFSS) (N = 464,664), and (3) the 2014 American Community Survey (ACS) conducted by the U.S. Census Bureau (N = 3.5 million per year). We predict asthma prevalence in the traditional survey (BRFSS) using social media information collected from Twitter and socioeconomic factors collected from ACS. The evidence suggests that monitoring asthma-related tweets may provide real-time information that can be used to predict outcomes from traditional surveys.

Keywords: asthma; social media monitoring; SMM; ACS; BRFSS; data linkage

Asthma is one of the most common chronic diseases, characterized by recurrent attacks of breathlessness and wheezing. In 2012, 18.7 million adults (1 in 12 adults) and 6.8 million

Hongying Dai is an associate professor in the Department of Health Services & Outcomes Research at Children's Mercy Hospital. Her research focuses on statistical epidemiology, social media monitoring, secondary analysis of big data, and development of novel statistical methods for medical research.

Brian R. Lee is an epidemiologist whose current research interests include antimicrobial stewardship, adverse drug reactions, health disparities, access to care, and clinical decision making. Lee currently serves as a senior biostatistician in Health Services & Outcomes Research at Children's Mercy Hospital.

Jianqiang Hao is vice president at the First National Bank of Omaha, where he manages the predictive analytics team. Hao is also an adjunct professor at the Bellevue University, and his research interests include big data analytics, social media data mining, and credit risk modeling.

Correspondence: hdai@cmh.edu

DOI: 10.1177/0002716216678399

children (1 in 11) in the United States had asthma (Centers for Disease Control and Prevention [CDC] 2015a) Asthma attack symptoms include coughing, wheezing, breathlessness, and chest tightness. Some common risk factors for asthma attacks are air pollution (Gowers et al. 2012; Tetreault et al. 2016), indoor and outdoor exposure to allergens (Platts-Mills and Carter 1997), tobacco smoke (McLeish and Zvolensky 2010), chemical irritants in the workplace (Tarlo 2000), psychosocial stress (Opolski and Wilson 2005), aspirin and medications, and physical exercise (Oddy et al. 2004; von Mutius et al. 2001). According to WHO estimates, 235 million people suffer from asthma globally. Although asthma cannot be cured, appropriate management can control the disorder and enable people to enjoy a good quality of life (Hoskins et al. 2000).

Asthma has a profound impact on people's lives and our society. Nearly half of asthmatic children misses at least one day of school each year due to asthma (Hsu et al. 2016) and nearly three in five adults with asthma limit their usual activities because of their asthma (CDC 2015a). Effective asthma care can be expensive, and many low-income people or racial minorities struggle to pay for the medicines and routine doctor visits they need. More than one in four black adults cannot afford their asthma medicines (CDC 2015a). Asthma costs the United States $56 billion each year and led to 479,300 hospitalizations, 1.9 million emergency department visits, and 8.9 million doctor visits in 2009. In 2008, asthma attributed to 10.5 million missed days of school and 14.2 million missed days of work (CDC 2015a). However, the traditional ways to monitor asthma are expensive and not very effective. A timely and accurate prediction of asthma prevalence can provide valuable information to healthcare providers for disease preparedness and has direct benefits for public health and hospital services.

Background

Current national asthma disease surveillance relies on annual surveys including The Behavioral Risk Factor Surveillance System (BRFSS) and the National Health and Nutrition Examination Survey, along with weekly reports to the Centers for Disease Control and Prevention (CDC 2015b). Traditional ways to collect asthma surveillance data are time consuming and costly to implement regularly. There are often long delays between data collection and information dissemination. Emerging social media monitoring (SMM) has bridged some of these gaps with three advantages for asthma or other public health surveillance: (1) Social media networks allow people to post back-stage information that might not be available from traditional surveys. Twitter provides a free-text platform of social networking and opinion expression. People may behave and express their feelings in a different way on Twitter compared with responses to traditional surveys; (2) Twitter provides researchers "big data" to analyze consumer attitudes and behaviors. An average of 500 million tweets are posted on Twitter each day, and 23 percent of online adults (19 percent of the entire adult population) currently use Twitter (Duggan et al. 2015). About 36 percent of Twitter users visit the site daily, and another 24 percent visit a few days a week. Facebook had 1.44 billion monthly active users as of the first quarter of 2015,[1] and 71 percent of adult Internet users currently use Facebook (Duggan et al. 2015). All these yield a large sample size and rich information for research, including sentiment assessment, prevalence by geographic location, and topics of asthma-related tweets; (3) Twitter information is real time and can be analyzed continuously with relatively little cost.

Social media networks provide researchers with "big data," automated and real-time information, to detect and monitor disease status. Prior studies have shown that social network sites could provide rich information to detect, monitor, and predict social events. A recent study by Ginsberg et al. (2009) demonstrated that Google search queries provide early and accurate detection of incidents of influenza reported by the CDC. Young, Rivers, and Lewis (2014) used geocoded tweets to evaluate the feasibility of using social networking data to detect and remotely monitor HIV outcomes. Their results show that there is a significant positive relationship between HIV-related tweets and HIV cases. Since social media communications are often through unstructured text data, their use imposes great challenges to data processing and analysis. However, text mining and sentiment analysis are useful in such analyses to extract important information from noisy data. Salathé and Khandewal (2011) assessed vaccination sentiments from 101,853 Twitter users and identified a strong correlation between sentiments expressed online and CDC-estimated vaccination rates by regions. Wong et al. (2015) estimated consumer sentiment related to the Affordable Care Act (Obamacare) using social media data from Twitter; they found that the estimated sentiment score is positively associated with enrollment in Obamacare at the state level. Dai and Hao (2016) developed a linkage using social media data and geographic information mapping along with the American Community Survey (ACS), and identified a significant geographic variation in e-cigarette opinion polarities, which were associated with social and economic factors.

However, most social media data are de-personalized and unstructured. It is a challenge to link social media data with other traditional data, such as survey and administrative data, psychometric and air quality data, and community indicators and social economic data. The main goal of our research is to link multiple large-scale data sources and construct an epidemiological model that accurately predicts asthma prevalence across regions. Such studies of asthma and its risk factors at the regional level have been limited. Geographical regions have distinctive characteristics in disease prevalence, socioeconomic status, and health outcomes. Characterization and identification of risk factors for asthma at the regional level will help health providers to improve health practices and prevent vulnerable communities from further deterioration. The objectives of this study are threefold: (1) extract real-time asthma-related information by mining conversations from Twitter; (2) link unstructured social media data with traditional structured data, including asthma prevalence from the CDC and socioeconomic variables from the U.S. Census; and (3) evaluate associations among asthma prevalence and public sentiment about asthma from SMM, socioeconomic conditions, and asthma prevalence at the U.S. state level as well as to develop a predictive model of asthma prevalence.

Data and Methods

This study analyzed data from three sources: asthma-related tweets from Twitter, sociodemographic data from the 2014 ACS, and asthma prevalence data from the 2014 BRFSS. All the data were linked through the geographic information system at the state level.

Social media monitoring (SMM)

Twitter, the fast-growing social networking company, has more than 400 million active users across the world, and more than 500 million tweets (messages that are fewer than 140 characters) are posted on Twitter each day. Twitter users can be followed by other twitter users, allowing others to receive and share tweets and thus distribute the tweets to a large audience quickly. It is estimated that Twitter is used by more men than women and by more young adults (18–49 years old) than older adults (50–65 years old) (Duggan et al. 2015).

Twitter provides public access to a random sample of approximately 1 percent of all tweets in real time through an Advanced Programming Interface (API) (Murthy 2013). R, a widely used data mining and statistics software, was used to access tweets and analyze the unstructured data in this study. We used the StreamR package to retrieve tweets and store the data in the collection of JSON (JavaScript Object Notation) files. The metadata includes the tweet text, time the tweet was sent, the user's language, number of friends and followers, the author's location, along with the geocode of latitude and longitude if the users chose to enable this feature. We further searched keywords such as "asthma," "wheezing," "sneezing," "inhaler," and "running nose" to collect tweets related to this study. These keywords were selected from a subset of search terms listed in a recent asthma study (Ram et al. 2015). Other potential keywords could be included to expand the search volume; however, there are tradeoffs to including a wide range of search terms. On one hand, increasing search terms can increase varying characteristics related to asthma. On the other hand, words have multiple meanings, and increasing a wide range of search terms may increase the number of tweets that are irrelevant to asthma. In this work, we decided to use highly relevant search terms that are commonly used to describe asthma, asthma symptoms, or the equipment treating asthma. Further text-mining techniques were applied to remove duplicate tweets and exclude non-English tweets. Twitter data were collected from July 2 to December 16, 2015 ($n = 168$ days) with a total of 1,254,710 unique tweets.

Usually, less than 1 percent of tweets have the geocode of longitude and latitude in the Twitter metadata; they do only if the user enabled this feature (Young, Rivers, and Lewis 2014). We further imputed the geographic locations

FIGURE 1
Study Design and Sample Sizes

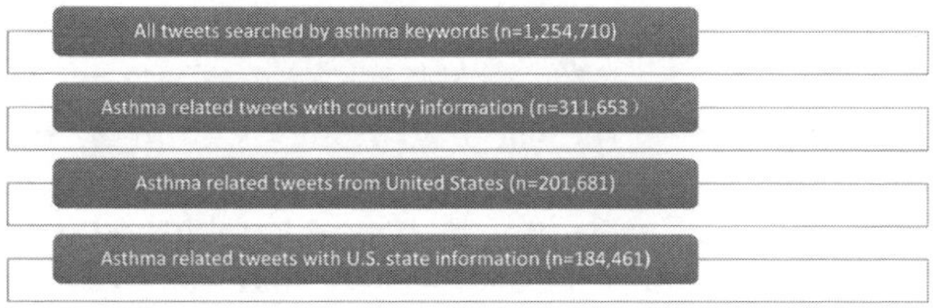

- All tweets searched by asthma keywords (n=1,254,710)
- Asthma related tweets with country information (n=311,653)
- Asthma related tweets from United States (n=201,681)
- Asthma related tweets with U.S. state information (n=184,461)

of tweeters based on their self-reported city, state, or country location in the Twitter metadata. With this approach, we were able to identify 24.8 percent of tweets with country information (311,653 / 1,254,710) and 93.4 percent of tweets with state information in the United States (188,461 / 201,681). Similar approaches have been used in other studies (Broniatowski et al. 2013; Daniulaityte et al. 2015; Ram et al. 2015). By expanding the location information from the users' public biographic profiles, we significantly increase the sample size, which enables us to analyze socioeconomic disparities. The sample sizes of asthma-related tweets with geographic locations are presented in Figure 1.

The prevalence of asthma tweets was calculated using the following formula:

$$\text{Prevalence of asthma tweets per 10,000 persons per day} = \frac{\text{number of asthma tweets in a state}}{\text{state population} \times \text{days of tweet collection} \times 1\%} \times 10,000,$$

where the state population as of July 1, 2014, is from the census data, tweet collection period is 168 days, and Twitter API data present approximately 1 percent of all real-time tweets.

As Twitter use varies by sociodemographic status, such as age, race/ethnicity, income, and education (Duggan et al. 2015), it is reasonable to assume that there might be regional variations in Twitter usage. We created another metric to measure the prevalence—we compared the asthma-related tweets with the overall tweets within the same region to calculate the relative index of asthma tweets (Young et al. 2014; Daniulaityte et al. 2015).

$$\text{Relative Index of Asthma Tweets} = \frac{\text{number of asthma tweets in a state} / 168}{\text{\# of overall tweets}}$$

Given the large number of tweets (nearly 500 million) posted on Twitter each day, we randomly collected tweets without using any search words from the Twitter API during the week of December 10 to December 16, 2015. To manage the search volume, we limited the search time to be two hours each day and randomly selected the time slot. We have a total of 300,000 random tweets with U.S. geolocation information.

Behavioral Risk Factor Surveillance System (BRFSS)

The BRFSS (CDC 2015b) is the nation's premier system of telephone surveys that collects data about U.S. residents' health-related risk behaviors, chronic health conditions, and use of preventive services. Established in 1984 within fifteen states, BRFSS now collects data in all fifty states as well as the District of Columbia and three U.S. territories. BRFSS completes more than 400,000 adult interviews each year, making it the largest continuously conducted health survey system in the world.

We extracted the adult self-reported current *asthma prevalence rate* from the 2014 BRFSS ($n = 464{,}664$). Two questions were asked in 2014, "Ever told you had asthma?" and "Do you still have asthma?" We calculated the asthma-ever ratio and asthma-now ratio at the state level to measure the percentage of the population that has ever had asthma and currently had asthma. Sample strata and weights were applied to account for the complex survey design and nonresponse. In addition, other candidate variables from the Adult Asthma Call-Back survey, such as asthma severity among adults with current asthma, overuse of quick-relief medication among persons with active asthma, and use of long-term control medication among persons with active asthma were also included in this study. These variables were treated as potential outcome variables.

American Community Survey (ACS)

The ACS (U.S. Census Bureau 2015) is a nationally representative survey conducted by the U.S. Census Bureau that regularly gathers socioeconomic information including *education, income, migration, disability, employment*, and *housing conditions*. The ACS is sent to approximately 295,000 addresses monthly (or 3.5 million per year), making it one of the largest ongoing surveys in the United States.

Previous studies have found disparities in asthma prevalence and asthma control based on multiple demographic and socioeconomic factors (Akinbami, LaFleur, and Schoendorfet 2002; Aligne et al. 2000; Gold and Wright 2005; Shaya et al. 2009; Williams, Sternthal, and Wright 2009). We selected four key variables from the 2014 ACS—age, race, education, and income—to investigate potential socioeconomic impacts on asthma prevalence and public sentiment and attitudes about asthma. These variables were used as covariates in the predictive model to control for potential confounding that might influence asthma prevalence. Some of these selected variables are: persons 65 years and older (%), July 1, 2014; non-Hispanic white population (%), July 1, 2014; bachelor's degree or higher, percent of persons age 25 years+, 2009–2013; median household income (in 2013 dollars), 2009–2013.

Sentiment score

Sentiment analysis is also called opinion mining (Liu 2012). It is the field of study that extracts and identifies the relevant emotions or opinions of a document concerning a specific event or interest. Unstructured data are common in sentiment analysis since people usually express their opinions through words instead of numbers. The sentiment for asthma, for example, can be represented by:

$$S_{ijk} = f(e_i, t_{ij}, h_{ijk}, X)$$

where e_i represents the ith event, such as asthma; t_{ij} represents the jth topic of the ith event, such as asthma symptoms, asthma diagnosis, and asthma treatments; h_{ijk} represents opinions from the kth individual regarding the jth topic of the ith event; and X represents other covariates that might affect sentiment, such as socioeconomic attributes.

We started with a preestablished opinion lexicon compiled by Liu,[2] which includes a list of positive and negative English sentiment words. The dataset includes 2,914 positive words and 4,914 negative words. Our exploratory analysis shows that positive words often do not provide information for asthma surveillance, thus only negative words, coded as negative values, were included in the final analysis and we also modified the general lexicon to better capture the opinions in context.

Sentiment score is calculated by summing the number of negative words within each tweet. The lower the sentiment score is, the more negatively people express their feelings.

$$Sentiment\ Score\ for\ each\ tweet = \sum_{i=1}^{n}(Negative\ word)$$

Analysis plan

In this study, we first analyzed the metadata collected from asthma-related tweets. We then compared the characteristics of tweets based on whether they were from the United States or other countries to identify any regional difference. Second, we analyzed the geographic variations of asthma-related tweets from the United States and examined whether the prevalence, relative index, and sentiment score of asthma-related tweets were different across the fifty states and the District of Columbia. Asthma-related tweets, along with their sentiment scores, and asthma prevalence from the BRFSS, and socioeconomic factors from the ACS were evaluated using the Spearman correlation. We chose this nonparametric test to prevent the bias from extreme values. The general linear model was used to develop a predictive model, where the dependent variable was the asthma prevalence rate from the BRFSS and explanatory variables were the asthma relative index and negative sentiment from Twitter, and socioeconomic variables from the ACS. Statistical analyses were performed using SAS 9.4 (Cary, NC) and a p-value of $< .05$ was considered statistically significant.

TABLE 1
Summary Statistics of Asthma-Related Tweets

	# of Tweets	# of Users	# of Tweets/ User	% of Tweets from Top 10 Users	% of Tweets with URLs	Potential Reach°	Average Reach/Tweet
Total	1,254,710	842,645	1.5	5.7%	19.3%	3,824,798,382	3,048
USA	201,681	126,661	1.6	8.7%	22.5%	809,795,803	4,015
Other countries	109,972	70,360	1.6	10.0%	24.5%	517,006,699	4,701

°Potential reach is measured by total times posted in Twitter, which were calculated by summing the total number of followers for each tweet.

Results

Summary of asthma-related tweets

A total of 1,254,710 asthma-related tweets were collected from July 2 to December 16, 2015 (168 days). Of all tweets, we were able to identify 311,653 tweets (24.8 percent) with country information and 201,681 (16.1 percent) were from the United States. A little more than 91 percent (91.4) of tweets in the United States (184,461/201,681) had state information derived from the tweeters' self-reported locations.

The summary statistics of Twitter metadata are presented in Table 1. These tweets were sent out by 842,645 unique Twitter users with 1.5 tweets per user. Approximately 6 percent of tweets were from the top ten users, and 19.3 percent of tweets had URLs. The potential reach from asthma-related tweets, calculated by summing the total number of followers for each tweet (Huang et al. 2014), was enormous (8,304,957 from approximately 1 percent of the Twitter API feed). When comparing the United States to other countries, the United States had the same number of tweets per user (1.6 vs. 1.6), lower percentage of tweets with URLs (22.5 percent vs. 24.5 percent), and lower average reach per tweet (4,015 vs. 4,701). A higher percentage of tweets from other countries (10.0 percent vs. 8.7 percent) were from the top ten users, indicating that some of these tweets might be sent through an automatic process on a bulk level.

The sentiment score was calculated at each tweet level with a mean of –64.5 and a standard deviation of 83.4. The sentiment score in the United States was lower than in the other countries (–69.8 vs. –64.6; $p < .0001$), indicating that American Twitter users might be more likely to express negative emotions about asthma compared with users from other countries. Figure 2 presents the trend of the daily average sentiment score and standard deviation from asthma-related tweets in the United States from July 2 to December 16, 2015.

Prevalence of asthma-related tweets

The prevalence of asthma-related tweets varied dramatically by state (Figure 3A). The top five states with the highest prevalence of asthma tweets were DC

FIGURE 2
The Daily Trend for Sentiment Score in the United States

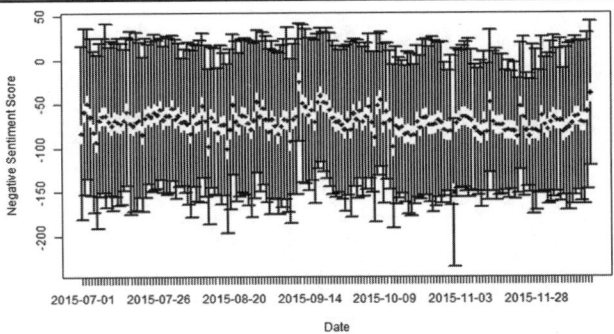

(29.1), TN (14.3), NH (11.0), MA (5.2), and NV (5.1); the five states with the lowest prevalence were WY (0.62), MS (0.63), WV (1.00), AR (1.16), and MT (1.18). The relative index of asthma-related tweets showed different patterns among U.S. states (Figure 3B). The five states with the highest relative index were HI (9.54), NH (8.30), TN (5.08), AK (3.31), and NM (1.97); the five states with the lowest relative index were KS (0.75), MN (0.82), KY (0.86), SC (0.89), and TX (0.91). It is worth noting that the ranking of prevalence and the relative index differed across U.S. states. For example, the District of Columbia had the highest prevalence of asthma-related tweets (29.3 per 10,000 persons), but the ranking dropped to the middle range after adjusting for total tweets (relative index: 1.41, ranked at 19th). A possible reason for this is that DC areas might have more Twitter users and thus have more asthma-related tweets, but the percentage of asthma tweets relative to other tweets was less prominent.

The sentiment score of asthma tweets also differed by state (Figure 3C). The five states with the highest sentiment score were WY (−50.8), MD (−58.8), MI (−60.3), ND (−60.6), and LA (−60.8); the five states with the lowest sentiment score were TN (−95.1), ME (−91.1), UT (−87.7), DE (−84.2) and MT (−83.3).

We further analyzed the correlations among asthma prevalence, relative index, and sentiment score from SMM and asthma statistics from the 2014 BRFSS. Both the relative index and sentiment score were significantly correlated with asthma statistics from the BRFSS. The relative index of asthma tweets was positively associated with both the asthma-ever ratio ($r = .32, p = .02$; Figure 4A) and the asthma-now ratio ($r = .38, p = .006$; Figure 4C), suggesting that a high asthma prevalence rate would drive more asthma-related tweets. Negatively significant associations were observed between sentiment score and asthma statistics from the BRFSS (asthma-ever, $r = -.30, p = .03$; Figure 4B; asthma-now, $r = -.28, p = .04$; Figure 4D), indicating that people were more likely to express negative feelings or experiences with asthma when the asthma prevalence rate was high. The relative index was also significantly correlated with LTCAdult (use of long-term control medication among adults) from the BRFSS ($r = .48, p = .002$; Figure 4E), suggesting that states with more people using long-term control

FIGURE 3
Choropleth Maps of Asthma Tweets and Sentiment Score in the United States

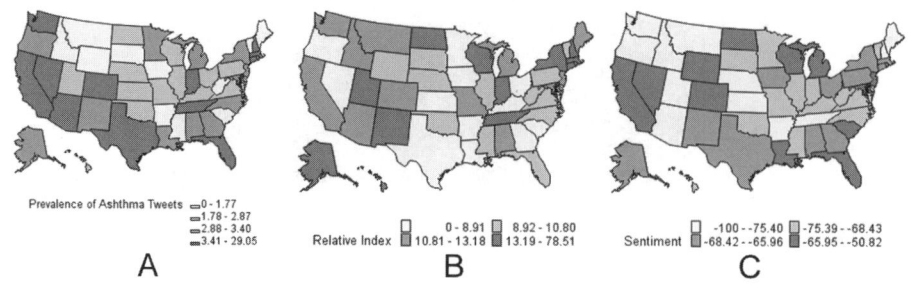

3A. Prevalence of asthma-related tweets
3B. Relative index of asthma tweets
3C. Negative sentiment score of asthma tweets
NOTE: States are divided into quartiles.

medication for asthma were more likely to tweet their experience or opinions. The asthma prevalence was not found to be significantly correlated with either the asthma-now ratio or the asthma-ever ratio.

Socioeconomic disparities

We evaluated the relationship of socioeconomic factors to the prevalence and relative index of asthma tweets. Asthma-related tweets were more prevalent in states with a higher median income (prevalence: $r = .37$, $p = .008$, Figure 5A; relative index: $r = .34$, $p = .02$, Figure 5B). A possible explanation is that the people with higher incomes were more likely to use Twitter, and they were more comfortable expressing their feelings and experience in social networks.

States with a higher percentage of residents with a bachelor's degree had a higher prevalence of asthma tweets ($r = .37$, $p = .008$, Figure 5C). This finding is consistent with the previous finding that Twitter users have higher education than the population at large (Duggan et al. 2015). States with a higher percentage of persons 65 years and older had lower sentiment scores ($r = -0.3$, $p = .03$, Figure 5D). This could be due to the fact that asthma might be more severe for older people; thus, they were more likely to express their negative feelings.

Predictive model

Two general linear models were developed to predict the asthma-ever rate from the BRFSS, using the sentiment score or relative index extracted from Twitter SMM and socioeconomic factors from the ACS.

$$Y = \alpha + \beta \times T + \theta \times X + \varepsilon$$

PREDICTING ASTHMA PREVALENCE

FIGURE 4
Scatterplots with Regression Line

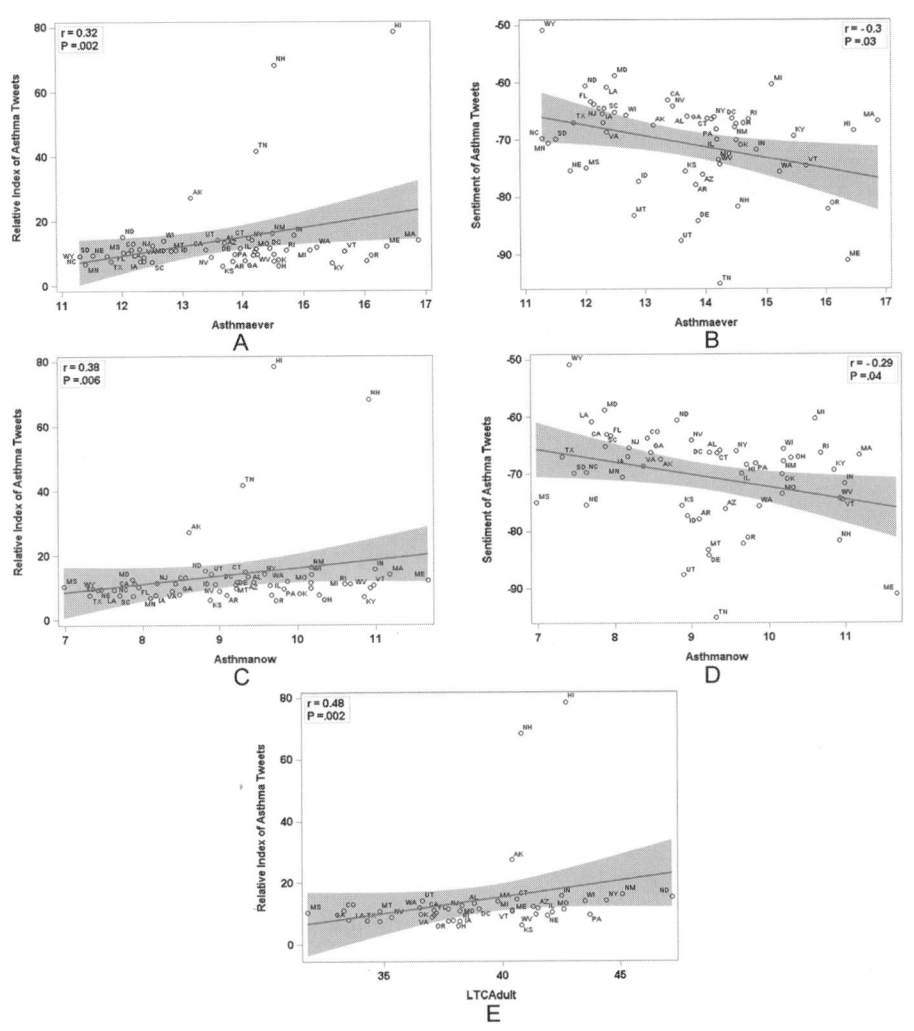

4A. Relative index of asthma tweets and asthma-ever ratio.
4B. Sentiment score of asthma tweets and asthma-ever ratio.
4C. Relative index of asthma tweets and asthma-now ratio.
4D. Sentiment score of asthma tweets and asthma-now ratio.
4E. Relative index of asthma tweets and use of long-term control medication among adults.

Where Y is the asthma-ever rate from BRFSS; T indicates the information extracted from Twitter, sentiment score, or relative index; and X includes a group of socioeconomic variables.

FIGURE 5
Scatterplots with Regression Line

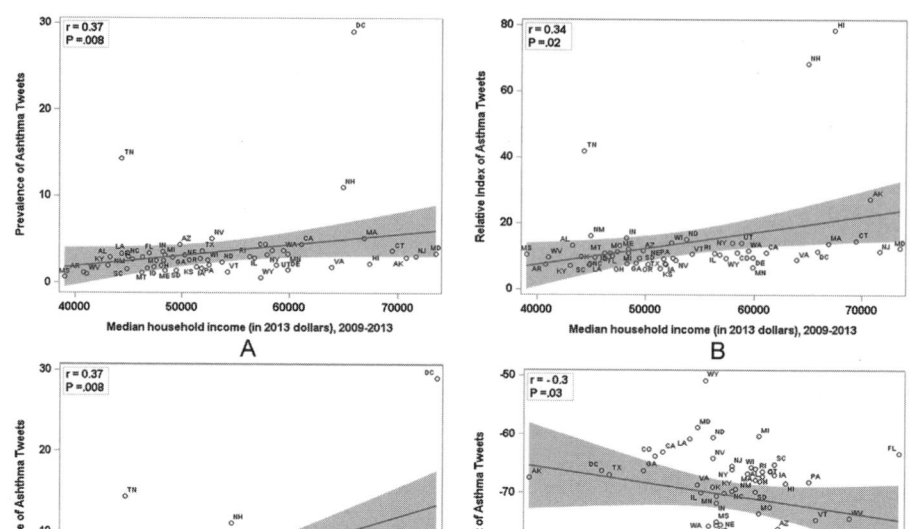

5A. Prevalence of asthma tweets with median household income.
5B. Relative index of asthma tweets and median household income.
5C. Prevalence of asthma tweets and percent of persons with bachelor's degree or higher.
5D. Sentiment score of asthma tweets and percent of persons with bachelor's degree or higher.

Table 2 presents the regression results in prediction of asthma-ever rate. After controlling for socioeconomic differences (i.e., age, race, education, and income), the models show that both negative sentiment score and relative index were significantly associated with the asthma-ever rate. The regression coefficient (−0.053 ± 0.024, $p = .03$) was negative for the sentiment score, indicating that a lower sentiment score is predictive of a higher asthma-ever rate. The regression coefficient (0.035 ± 0.016, $p = .03$) was positive for the relative index, indicating that a higher relative index is predictive of a higher asthma-ever rate. Another significant covariate was the percentage of persons 65 years and older (0.035 ± 0.016, $p = .04$), which was associated with a higher prevalence of asthma-ever.

Similar models were developed to predict asthma-now prevalence (Table 3). After controlling for other covariates, the regression coefficient for the sentiment score (−0.031 ± 0.019, $p = .12$) was negative but not significant. The regression coefficient for the relative index (0.031 ± 0.013, $p = .02$) continued to be positive and significant. In both models, the percentage of persons 65 years and older was significantly associated with a higher prevalence of asthma-now.

TABLE 2
Predictive Model for Asthma-Ever Prevalence from the BRFSS

Model I

Parameters	Regression coefficient (estimate ± standard error)	p value
Intercept	5.586 ± 3.007	.07
Sentiment score from Twitter	−0.058 ± 0.024	.02
Persons 65 years and over (%), ACS	0.240 ± 0.114	.04
Percent white only, ACS	−0.017 ± 0.015	.27
Bachelor degree or higher (%), ACS	0.038 ± 0.050	.45
Median household income (1,000 $), ACS	0.010 ± 0.040	.74

Model II

Parameters	Regression coefficient (estimate ± standard error)	p value
Intercept	9.610 ± 2.687	< .01
Relative index from Twitter	0.036 ± 0.017	.04
Persons 65 years and over (%), ACS	0.198 ± 0.119	.10
Percent white only, ACS	0.002 ± 0.016	.90
Bachelor degree or higher (%), ACS	0.085 ± 0.052	.11
Median household income (1,000 $), ACS	−0.040 ± 0.040	.33

Discussion

Summary

Asthma is one of the common chronic diseases with a profound impact on people's lives. It also has a high economic burden to our society through direct expenses and indirect lost time and productivity. In this study, we developed an epidemiological model to predict the regional level of asthma prevalence by linking multiple data sources—emerging SMM data with unstructured, free-text information and traditional surveys (BRFSS and ACS data).

Social media, with high reach and quick dissemination, is an important venue to collect asthma-related information. The recent rise of "big data" analytics offers a great opportunity to measure the perception of asthma among a large population. People may behave and express their feelings in a different way on social media networks than they do when they respond to traditional surveys. In addition, the opinion information extracted from unstructured, free-text messages from SMM can be monitored in real time. Therefore, social media posts may contain "back-stage" information that might not be available from a traditional survey. In this work, we identified a proliferation of tweets related to asthma along with a high potential reach for these tweets (1,254,710 tweets from

TABLE 3
Predictive Model for Asthma-Now Prevalence from the BRFSS

Model I

Parameters	Regression coefficient (estimate ± standard error)	p value
Intercept	0.978 ± 2.44	.69
Sentiment score from Twitter	−0.031 ± 0.019	.12
Persons 65 years and over (%), ACS	0.235 ± 0.092	.01
Percent white only, ACS	0.017 ± 0.013	.19
Bachelor degree or higher (%), ACS	0.031 ± 0.041	.45
Median household income (1,000 $), ACS	0.010 ± 0.030	.79

Model II

Parameters	Regression Coefficient (estimate ± standard error)	p value
Intercept	3.273 ± 2.080	.12
Relative index from Twitter	0.031 ± 0.013	.02
Persons 65 years and over (%), ACS	0.189 ± 0.092	.04
Percent white only, ACS	0.031 ± 0.013	.02
Bachelor degree or higher (%), ACS	0.065 ± 0.040	.11
Median household income (1,000$), ACS	−0.030 ± 0.030	.32

July 2 to December 16, 2015, with a potential reach of 22,766,657 per day from approximately 1 percent of tweets).

Our analyses show that real-time asthma-related information could be effectively extracted from SMM. The relative index of asthma-related tweets was significantly predictive of the asthma prevalence rate from the BRFSS (0.035 ± 0.016, p = .03). The sentiment score developed by text mining the tweets was also a significant predictor of the asthma ever rate (−0.053 ± 0.024, p = .03) in the general linear model. Our study identified regional variation of sentiment, prevalence, and relative index of asthma tweets and their associations with socioeconomic status, which suggests that some regions may suffer from a higher burden of asthma. Infoveillance, the continuous monitoring of public messages regarding asthma through social media, is essential to parse out these, and other, differences related to asthma prevalence.

Limitations

Our study is not without limitations. First, our research was based on a sample of Twitter users with selected search keywords related to asthma. A previous study has shown that Twitter users are more skewed to members of minority groups (27 percent of African American online adults and 25 percent of Hispanic online adults, compared with 21 percent of white online adults) and younger generations (37 percent of 18–29 year-old online adults, compared with 12 percent of 54–64 year-old online adults) (Duggan et al. 2015). More studies should be conducted to better understand the profile of social media users, compare them with traditional survey populations, and use the findings to make inferences about the general population. Furthermore, we used only a limited set of keywords to collect a large number of asthma-related tweets with relatively high relevance; we might have missed some asthma-related tweets. For example, Gillingham et al. (2013) included names of prescription drugs used to treat the condition, such as "albuterol" and "Singulair" in their keyword search. Even in our current selected keywords, "sneezing" and "wheezing" are not unique symptoms of asthma patients; thus, we might have included some tweets unrelated to asthma. In addition, we used the self-reported metadata to extract the geolocation information. Self-reported information might be inaccurate (Hecht et al. 2011) or not reflect the users' current location. Continuous improvement of geocoding and information extraction techniques could help to mitigate these limitations in future studies.

Second, one of the challenges in working with social media data is the amount of "noise" or "chatter" (misinformation) included in the data. In our study, nearly 6 percent of asthma-related tweets were from the top ten users, and 19.3 percent had URLs, suggesting that some of these tweets might be sent through twitter bots or spammers (Benevenuto et al. 2010). Some earlier studies (Salathé and Khandelwal 2011; Dai and Hao 2016) used a combination of human judgement and machine learning algorithms to classify the tweets and remove the unrelated tweets from the study. Similar methods could be used in asthma surveillance to identify the most relevant tweets.

Third, this study used a predefined lexicon (Liu 2012) to extract the sentiment associated with each tweet. Although we have tried to modify this lexicon to better capture the opinions from asthma-related tweets, the true emotions or feelings might not be accurately reflected in the calculated sentiment score given the complexity of human language. In addition, we included only negative words in the study without analyzing the positive or neutral words. A more sophisticated model that reflects different topics of asthma tweets, such as asthma symptoms, asthma diagnosis, and asthma treatment, might provide additional insights for prediction of asthma prevalence.

Finally, we linked only the Twitter data from SMM and sociodemographic data from the ACS to examine the association with asthma prevalence in this study. Other sources of data, such as Google Trends and emergency department visits, could also provide valuable insights into health. For example, Google Trends data, an index that shows how often a particular term is searched in Google relative to the total search volume across various regions and time horizons, has proven to be an effective predictor in influenza detection (Ginsberg et al. 2009), private consumption (Vosen and Schmidt 2011), disease tracking (Pelat et al. 2009), electronic cigarette use (Ayers, Ribisl, and Brownsteinet 2011; Ayers et al. 2016), and other public health insights (Nuti et al. 2014). Air pollution has also been found to be correlated with asthma prevalence and severity (Gowers et al. 2012; Schwartz et al. 1993; Tetreault et al. 2016; Tolbert et al. 2000; Whittemore and Korn 1980). Future research could link the multilevel data from Twitter SMM, Google Trends, BRFSS, Air Quality, and emergency department visits through the geographic information system and develop time series models to forecast daily emergency department visits and asthma prevalence.

Conclusion

Our study affirms that social media data have the potential to serve as a rapid, cost-effective health detection system with real-time information to monitor chronic disease and track public sentiment. The characterization and identification of risk factors for asthma at the community level could also help health providers to improve health practices and prevent vulnerable communities from further deterioration. We have laid the groundwork by using sentiment and a relative prevalence index to predict asthma prevalence from traditional surveys. Our methodology can be directly extended to county or other smaller geographic region level, which would provide a granular community profile. In this study, we did not link data at the county level due to data limitations. Twitter API provided a free sample of 1 percent randomly selected tweets, and less than 0.3 percent (around 3,000) of tweets had geolocation in our study, which made the sample size for the county-level analysis too small. Our current mapping method relies on the self-reported geolocation provided by Twitter users; such information becomes less accurate beyond the state or large metropolitan area. Future research could study a full set of Twitter data, which could be purchased from commercial vendors, and perform the analysis at the county level or smaller geographic region level. Understanding public sentiment and the prevalence of asthma-related information will assist researchers and public health departments in collecting supplementary information regarding attitudes and experience related to asthma.

Notes

1. Statista. Number of monthly active Facebook users worldwide as of 1st quarter 2015; see http://www.statista.com/statistics/264810/number-of-monthly-active-facebook-users-worldwide/.

2. See Opinion Lexicon (or Sentiment Lexicon) section at https://www.cs.uic.edu/~liub/FBS/sentiment-analysis.html#lexicon.

References

Akinbami, Laura J., Bonnie J. LaFleur, and Kenneth C. Schoendorf. 2002. Racial and income disparities in childhood asthma in the United States. *Ambulatory Pediatrics* 2 (5): 382–87.

Aligne, Andrew C., Peggy Auinger, Robert S. Byrd, and Michael Weitzman. 2000. Risk factors for pediatric asthma. Contributions of poverty, race, and urban residence. *American Journal of Respiratory and Critical Care Medicine* 162 (3 Pt 1): 873–77.

Ayers, John W., Benjamin M. Althouse, Jon-Patrick Allem, Eric C. Leas, Mark Dredze, and Rebecca S. Williams. 2016. Revisiting the rise of electronic nicotine delivery systems using search query surveillance. *American Journal of Preventive Medicine* 50 (6): e173–e181.

Ayers, John W., Kurt M. Ribisl, and John S. Brownstein. 2011. Tracking the rise in popularity of electronic nicotine delivery systems (electronic cigarettes) using search query surveillance. *American Journal of Preventive Medicine* 40 (4): 448–53.

Benevenuto, Fabrício, Gabriel Magno, Tiago Rodrigues, and Virgílio Almeida. 2010. Detecting spammers on Twitter. *(CEAS 2010) Seventh Annual Collaboration, Electronic messaging, Anti-Abuse and Spam Conference*. Available from http://www.decom.ufop.br/fabricio/download/ceas10.pdf.

Broniatowski, David A., Michael J. Paul, and Mark Dredze. 2013. National and local influenza surveillance through Twitter: An analysis of the 2012–2013 influenza epidemic. *PLoS One* 8 (12): e83672.

Centers for Disease Control and Prevention (CDC). 2015a. *Asthma's impact on the nation: Data from the CDC National Asthma Control Program*. Washington, DC: CDC. Available from http://www.cdc.gov/asthma/impacts_nation/asthmafactsheet.pdf.

Centers for Disease Control and Prevention (CDC). 2015b. Behavioral risk factor surveillance system. Available from http://www.cdc.gov/brfss/.

Dai, Hongying, and Jianqiang Hao. 2016. Mining social media data for opinion polarities about electronic cigarettes. *Tobacco Control*. doi:10.1136/tobaccocontrol-2015-052818

Daniulaityte, Raminta, Ramzi W. Nahhas, Sanjaya Wijeratne, Robert G. Carlson, Francois R. Lamy, Silvia S. Martins, Edward W. Boyer, G. Alan Smith, and Amit Sheth. 2015. "Time for dabs": Analyzing Twitter data on marijuana concentrates across the U.S. *Drug and Alcohol Dependence* 155:307–11.

Duggan, Maeve, Nicole B. Ellison, Cliff Lampe, Amanda Lenhart, and Marry Madden. 2015. Social media update 2014. Pew Research Center. Available from http://www.pewinternet.org/2015/01/09/social-media-update-2014/.

Gillingham, Gwendolyn, Michael A. Conway, Wendy W. Chapman, Michael B. Casale, and Kathryn B. Pettigrew. 2013. #wheezing: A content analysis of asthma-related tweets. *Online Journal of Public Health Informatics* 5 (1). doi:10.5210/ojphi.v5i1.4591.

Ginsberg, Jeremy, Mattew H. Mohebbi, Rajan S. Patel, Lynnette Brammer, Mark S. Smolinski, and Larry Brilliant. 2009. Detecting influenza epidemics using search engine query data. *Nature* 457 (7232): 1012–14.

Gold, Diane R., and Rosalind Wright. 2005. Population disparities in asthma. *Annual Review of Public Health* 26:89–113.

Gowers, Alison M., Paul Cullinan, Jon G. Ayres, H. Ross Anderson, David P. Strachan, Stephen T. Holgate, Inga C. Mills, and Robert L. Maynard. 2012. Does outdoor air pollution induce new cases of asthma? Biological plausibility and evidence; a review. *Respirology* 17 (6): 887–98.

Hecht, Brent, Lichan Hong, Bongwon Suh, and Ed H. Chi. 2011. Tweets from Justin Bieber's heart: The dynamics of the "location" field in user profiles. *CHI 2011*. Available from https://pdfs.semanticscholar.org/c710/163721ed4c4587f4308951033bb712d57515.pdf.

Hoskins, G., C. McCowan, R. G. Neville, G. E. Thomas, B. Smith, and S. Silverman. 2000. Risk factors and costs associated with an asthma attack. *Thorax* 55 (1): 19–24.

Hsu, Joy, Xiaoting Qin, Suzanne F. Beavers, and Maria C. Mirabelli. 2016. Asthma-related school absenteeism, morbidity, and modifiable factors. *American Journal of Preventive Medicine* 51 (1): 23–32.

Huang Jidong, Rachel Kornfield, Glen Szczypka, and Sherry Emery. 2014. A cross-sectional examination of marketing of electronic cigarettes on Twitter. *Tobacco Control* 23 (Suppl 3): iii26–iii30.

Liu, Bin. 2012. *Sentiment analysis and opinion mining*. San Rafael, CA: Morgan & Claypool Publishers.

McLeish, Alison C., and Michael J. Zvolensky. 2010. Asthma and cigarette smoking: A review of the empirical literature. *Journal of Asthma* 47 (4): 345–61.

Murthy, Dhiraj. 2013. *Twitter: Social communication in the Twitter age*. Cambridge: Polity.

Nuti, Sudhakar V., Brian Wayda, Isuru Ranasinghe, Sisi Wang, Rachel P. Dreyer, Serene I. Chen, and Karthik Murugiah. 2014. The use of Google trends in health care research: A systematic review. *PLoS One* 9 (10): e109583.

Oddy, Wendy H., Jill L. Sherriff, Nicholas H. de Klerk, Garth E. Kendall, Peter D. Sly, Lawrence J. Beilin, Kevin B. Blake, Louis I. Landau, and Fiona J. Stanley. 2004. The relation of breastfeeding and body mass index to asthma and atopy in children: A prospective cohort study to age 6 years. *American Journal of Public Health* 94 (9): 1531–37.

Opolski, Melissa, and Ian Wilson. 2005. Asthma and depression: A pragmatic review of the literature and recommendations for future research. *Clinical Practice & Epidemiology in Mental Health*. doi:10.1186/1745-0179-1-18

Pelat, Camille, Clement Turbelin, Avner Bar-Hen, Antoine Flahault, and Alain-Jacques Valleron. 2009. More diseases tracked by using Google Trends. Letter to the editor. *Emerging Infectious Disease* 15 (8): 1327–28.

Platts-Mills, Thomas A. E., and Melody C. Carter. 1997. Asthma and indoor exposure to allergens. *New England Journal of Medicine* 336 (19): 1382–84.

Ram, Sudha, Wenli Zhang, Max Williams, and Yolande Pengetnze. 2015. Predicting asthma-related emergency department visits using big data. *IEEE Journal of Biomedical Health Informatics* 19 (4): 1216–23.

Salathé, Marcel, and Shashank Khandelwal. 2011. Assessing vaccination sentiments with online social media: Implications for infectious disease dynamics and control. *PLoS Computational Biology* 7 (10): e1002199.

Schwartz, Jeol, Daniel Slater, Timothy V. Larson, William E. Pierson, and Jane Q. Koenig. 1993. Particulate air pollution and hospital emergency room visits for asthma in Seattle. *American Review of Respiratory Disease* 147 (4): 826–31.

Shaya, Fadia T., Mark S. Maneval, Confidence M. Gbarayor, Kyongsei Sohn, Anand A. Dalal, Dongyi Du, and Steven M. Scharf. 2009. Burden of COPD, asthma, and concomitant COPD and asthma among adults: Racial disparities in a Medicaid population. *Chest* 136 (2): 405–11.

Tarlo, S. M. 2000. Workplace respiratory irritants and asthma. *Occupational Medicine* 15 (2): 471–84.

Tetreault, Louis-Francois, Marieve Doucet, Philippe Gamache, Michel Fournier, Allan Brand, Tom Kosatsky, and Audrey Smargiassi. 2016. Childhood exposure to ambient air pollutants and the onset of asthma: An administrative cohort study in Quebec. *Environmental Health Perspectives* 124 (8): 1276–82.

Tolbert, Paige E., James A. Mulholland, David L. MacIntosh, Fan Xu, Danni Daniels, Owen J. Devine, Bradley P. Carlin, Mitchel Klein, Andre J. Butler, Dale F. Nordenberg, Howard Frumkin, P. Barry Ryan, and Macy C. White. 2000. Air quality and pediatric emergency room visits for asthma in Atlanta, Georgia, USA. *American Journal of Epidemiology* 151 (8): 798–810.

U.S. Census Bureau. 2015. American Community Survey (ACS). Available from https://www.census.gov/programs-surveys/acs/.

von Mutius, E., J. Schwartz, L. M. Neas, D. Dockery, and S. T. Weiss. 2001. Relation of body mass index to asthma and atopy in children: The National Health and Nutrition Examination Study III. *Thorax* 56 (11): 835–38.

Vosen, Simeon, and Torsten Schmidt. 2011. Forecasting private consumption: survey-based indicators vs. Google trends. *Journal of Forecasting* 30 (6): 565–78.

Whittemore, A. S., and E. L. Korn. 1980. Asthma and air pollution in the Los Angeles area. *American Journal of Public Health* 70 (7): 687–96.

Williams, David R., Michelle Sternthal, and Rosalind J. Wright. 2009. Social determinants: Taking the social context of asthma seriously. *Pediatrics* 123 (Suppl 3): S174–S184.

Wong, Charlene A., Maarten Sap, Andrew Schwartz, Robert Town, Tom Baker, Lyle Ungar, and Raina M. Merchant. 2015. Twitter sentiment predicts Affordable Care Act marketplace enrollment. *Journal of Medical Internet Research* 17 (2):e51.

Young, Sean D., Caitlin Rivers, and Bryan Lewis. 2014. Methods of using real-time social media technologies for detection and remote monitoring of HIV outcomes. *Preventive Medicine* 63:112–15.

Correlates of Contraceptive Use and Health Facility Choice among Young Women in Malawi

By
JEAN DIGITALE,
STEPHANIE PSAKI,
ERICA SOLER-HAMPEJSEK,
and
BARBARA S. MENSCH

We explore whether differential access to family-planning services and the quality of those services explain variability in uptake of contraception among young women in Malawi. We accomplish this by linking the Malawi Schooling and Adolescent Study, a longitudinal survey of young people, with the Malawi Service Provision Assessment collected in 2013–14. We also identify factors that determine choice of facility among those who use contraception. We find that the presence and characteristics of nearby facilities with contraception available did not appear to affect use. Rather, characteristics such as facility type and whether contraception was provided free of charge determined where women deciding to use contraception obtained their contraception. We argue that in a context where almost all respondents resided within 10 kilometers of a health facility, improving access to and quality of family-planning services may not markedly increase contraceptive use among young women without broader shifts in norms regarding childbearing in the early years of marriage.

Keywords: Malawi; contraception; access; family-planning; adolescents

Although fertility has declined in sub-Saharan Africa, the region continues to have the highest rates of childbearing and population growth in the world (United Nations Department of Economic and Social Affairs Population Division

Jean Digitale is a data analyst/research coordinator at the Population Council. She has contributed to cleaning and analysis of data from several studies, including the longitudinal Malawi Schooling and Adolescent Study, which draws on in-depth data from more than 2,500 adolescents to elucidate relationships among young people's schooling experiences, learning, and health outcomes.

Stephanie Psaki is an associate in the Poverty, Gender, and Youth Program at the Population Council. She conducts research on girls' education, sexual and reproductive health, and violence. Psaki also serves as the editor of Studies in Family-Planning, *a scholarly journal published by the Council.*

Correspondence: jdigitale@popcouncil.org

DOI: 10.1177/0002716216678591

2015). Indeed, fertility declines in sub-Saharan Africa appear to be slower than those in Asia and Latin America at comparable stages in their fertility transitions (Bongaarts and Casterline 2012). Additionally, unmet need for contraception, as conventionally measured, is higher in Africa than in other regions; a relatively small percentage of those who report that they do not want to become pregnant are using a contraceptive method (Bongaarts and Casterline 2012; Sedgh and Hussain 2014).

Whether family-planning services or investments in other aspects of development, such as women's education, contribute to childbearing declines in high fertility countries is an ongoing and unresolved debate in the demographic literature (Ainsworth, Beegle, and Nyamete 1996; Demeny 1992; Jain and Ross 2012; Lutz 2014; Pritchett 1994). Those who have argued that increased use of contraception results from reduced demand for children advocate investment in economic and social development (Ainsworth, Beegle, and Nyamete 1996; Casterline and Sinding 2000; Demeny 1992; Lutz 2014; Pritchett 1994). Pritchett (1994) claims that fertility desires, not contraceptive access, are critical to achieving reductions in fertility. A recent study, controlling for the endogeneity of education, found that with increased schooling, desired fertility fell in Malawi, Uganda, and Ethiopia—evidence that additional years of schooling indeed reduces preferences for a large number of children (Behrman 2015). Those who have focused their research on describing and assessing family-planning services assert that improving the accessibility, availability, and quality of these services will increase demand for and use of contraception (Magnani et al. 1999; RamaRao et al. 2003; Ross and Hardee 2013; Ross and Stover 2013; Skiles et al. 2015; Wang et al. 2012; Yao, Murray, and Agadjanian 2013) by converting some fraction of those with unmet need into users (Casterline and Sinding 2000). Although inadequate access to services has not been shown to be a primary contributor to unmet need (Bongaarts and Bruce 1995; Casterline and Sinding 2000; Choi, Fabic, and Adetunji 2016; Sedgh and Hussain 2014), research focusing on the service environment has found that better quality of care appears to be associated with

---

Erica Soler-Hampejsek is an associate at the Population Council. She uses various statistical methods to study the influence of schooling on the timing of sexual initiation, marriage, and childbearing. She has collected and analyzed longitudinal data in Guatemala, Malawi, and Zambia, and directed five rounds of data collection for the Malawi Schooling and Adolescent Survey.

Barbara Mensch is a senior associate at the Population Council. She has conducted research on the quality of family-planning services in developing countries and its effect on contraceptive use, transitions to adulthood in Africa, reliability of self-reports relating to sexual behavior in demographic surveys, and the behavior and characteristics of HIV prevention trial participants.

NOTE: This research was funded by the Eunice Kennedy Shriver National Institute of Child Health and Human Development (R01 HD047764 and R01 HD062155) and the Spencer, John D. and Catherine T. MacArthur, and William and Flora Hewlett Foundations. Content is the sole responsibility of the authors and does not necessarily represent the official views of the funding institutions. The authors would like to acknowledge John Bongaarts, Clara Burgert, Margaret Frye, Sandra Hofferth, Mark Montgomery, Wenjuan Wang, and our anonymous reviewers for their helpful comments.

greater family-planning uptake and continuity of use (Jain 1989; Koenig, Hossain and Whittaker 1997; Mensch, Arends-Kuenning and Jain 1996; RamaRao et al. 2003).

With approximately 60 percent of the population in sub-Saharan Africa under the age of 25, even if the total fertility rate (TFR) decreases substantially in the near future, the region will still be on a trajectory to experience rapid population growth as a result of population momentum (Bongaarts 1994; United Nations Department of Economic and Social Affairs Population Division 2015). However, because differing paces of fertility decline would lead to vast differences in future population size (Casterline 2001), decisions about family-planning made by the current cohort of youth will play a key role in determining population size in the future.

In this article, we address gaps in the existing literature on determinants of contraceptive use in sub-Saharan Africa by focusing on the health facility choices of young women. Previous analyses attempting to quantify the effect of availability of family-planning services on fertility have typically focused on all reproductive-aged women (Ainsworth, Beegle, and Nyamete 1996; Do and Kurimoto 2012; Jain and Ross 2012; Skiles et al. 2015; Yao, Murray, and Agadjanian 2013), despite the fact that younger women may have different attitudes toward contraceptive use and different experiences interacting with family-planning providers than do older women. Further, little has been written about determinants of health facility choice, and the existing literature is somewhat contradictory. For example, Akin and Rous (1997) found that increased distance and having other services available at a facility were deterrents to choice of that facility in the Philippines, but there was no evidence that other provider characteristics were associated with facility choice. A study in Burkina Faso, Ghana, Malawi, and Uganda found that adolescents had positive views about public clinics, where they perceived confidentiality and accessibility to be high and cost low, and preferred these clinics to private clinics, likely because of cost (Biddlecom et al. 2007). In contrast, in an analysis comparing public and private health facilities in Tanzania, Kenya, and Ghana, client satisfaction was higher at private facilities than at public facilities, likely due to "factors such as shorter waiting times and fewer stockouts of methods and supplies" (Hutchinson, Do, and Agha 2011). Understanding the determinants of health facility choice is important because these factors—distance, availability of contraceptive methods, cost, and so on—may influence whether young women initiate and continue use of contraception throughout their reproductive lives.

To address this, we link data from the Malawi Schooling and Adolescent Study—a longitudinal survey of young people first interviewed in 2007 when they were between 14 and 17 years old and last interviewed in 2013—with data from the Malawi Service Provision Assessment (Ministry of Health [Malawi] and ICF International 2014) collected in 2013–2014. We explore whether differential access to family-planning services explains variability in uptake of contraception among young women and identify the factors that determine choice of facility among those who use contraception.

Background

Family-planning needs of young women

More than three-quarters of females in Eastern and Southern Africa are estimated to have had at least one sexual experience by the age of 20 (Lloyd 2005). Levels of unmet need—typically defined as nonuse of contraception among those who do not want to become pregnant—are reported to be very high (40 percent or more) among sexually active, unmarried young women in more than half the countries in the region. Levels of both contraceptive use and unmet need are higher among sexually active, unmarried young women than among currently married young women in most of the region (Khan and Mishra 2008), likely due to pressure to conceive shortly after marriage (Hindin and Fatusi 2009). Motherhood is a fundamental element of married women's identity and enhances social status (Cooper et al. 2007) in sub-Saharan Africa; being childless is generally regarded as undesirable (Dyer 2007).

Family-planning programs in sub-Saharan Africa have been criticized for primarily targeting those who are married or who have been pregnant at least once, ignoring the needs of unmarried sexually active youth (Prata, Weidert, and Sreenivas 2013). The timing of contraceptive use relative to first birth and marriage is an important consideration in determining which women the family-planning programs are reaching (Defo 2011; Garenne, Tollman, and Kahn 2000; Prata, Weidert, and Sreenivas 2013). Reasons cited for unmarried adolescents not seeking family-planning services include embarrassment or fear, cost, and lack of knowledge about where to obtain contraception (Biddlecom et al. 2007). Although practitioners have sought to develop "youth-friendly" services to address these barriers, based on current levels of contraceptive use among unmarried young women who report being sexually active, this approach has shown limited success thus far (Prata, Weidert, and Sreenivas 2013).

Fertility in Malawi

Malawi's demographic profile is fairly typical of sub-Saharan Africa, with 45 percent of the population under the age of 15 (Population Reference Bureau 2014). The TFR in Malawi fell from 6.7 births per woman in 1992 (National Statistical Office and Macro International Inc. 1994), to 6.3 in 2000 (National Statistical Office [Malawi] and ORC Macro 2001), 6.0 in 2004 (National Statistical Office [Malawi] and ORC Macro 2005), 5.7 in 2010 (National Statistical Office and ICF Macro 2011), and 4.4 as of 2016 (National Statistical Office and ICF International 2016). Among countries in sub-Saharan Africa, Malawi has made considerable progress in increasing uptake of modern contraceptives among those who desire to limit births (Sharan et al. 2011; USAID/Africa Bureau et al. 2012). In Eastern Africa, the contraceptive prevalence rate, including modern and traditional methods, among women aged 15–49 years old who were married or in a union was 12 percent in 1990 and increased to 33 percent by 2010. The contraceptive prevalence rate in Malawi among married women was at approximately the same 12 percent regional average in 1990, but increased to 45 percent by 2010 (Alkema et al. 2013). As of 2016, use of modern contraception among currently married women in Malawi has reached 58 percent (and among sexually active unmarried women it is now 43 percent; National Statistical Office and ICF International 2016). The large increase observed in contraceptive prevalence between 2004 and 2010 did not lead to a commensurate reduction in the TFR (Jain et al. 2014), although TFR has decreased further as of 2016. While unmet need among women in Malawi aged 15–49 years old who were married or in a union decreased from 36 percent in 1992 (National Statistical Office and Macro International Inc. 1994) to 19 percent in 2016, among sexually active unmarried women, unmet need remains high (40 percent; National Statistical Office and ICF International 2016). A consequence of this consistently high level of unmet need is that close to half of pregnancies in Malawi are considered mistimed or unwanted (National Statistical Office and ICF Macro 2011); this is among the highest levels in sub-Saharan Africa (Johnson, Abderrahim, and Rutstein 2011). Additionally, although the proportion of adolescents aged 15–19 who have begun childbearing decreased from 2004 (34 percent) to 2010 (26 percent), it then increased again to 29 percent in 2016 (National Statistical Office and ICF International 2016).

There are considerable regional differences within Malawi. Young women in the southern region, where the Malawi Schooling and Adolescent Study sample resided at baseline, tend to have poorer sexual and reproductive health outcomes than those in the rest of the country. They have a lower median age at first birth (18.5 among 20–24-year-olds) than those in the northern and central regions (19.1 and 19.2, respectively). HIV prevalence among women age 15–49 is highest in the southern region (18 percent) compared with 8 percent in the northern region and 9 percent in the central region (National Statistical Office and ICF Macro 2011).[1] Sociocultural and religious differences exist between regions as well, which likely influence fertility. Ethnic groups in the North are patrilineal, whereas the Yao tribe and others in the South are matrilineal. Eighty-three percent of Malawians are Christian (Malawi National Statistical Office 2008); however, three-quarters of the Yao tribe are Muslim and maintain Yao traditions along with Islamic ones (Chimbiri 2006).

Family-planning policies and programs in Malawi

The rapid expansion of contraceptive use in Malawi reflects, in part, the apparent commitment on the part of the government to reduce fertility through expansion of the national family-planning program (Respond Project 2012). Both the Maputo Plan of Action in 2006 and a policy analysis by the Malawi Ministry of Development Planning and Cooperation in 2010 recognized that rapid population growth might outpace economic growth and the government's ability to provide social services (African Union Commission 2006; Ministry of Development Planning and Cooperation 2010). In 2009, Malawi's government "domesticated" the Maputo Plan of Action via the development of the Sexual and Reproductive Health and Rights Policy, intended to guide program managers of government health departments; nongovernmental, community, and faith-based organizations; and the private sector to effectively develop sexual and reproductive health services responsive to the needs of the Malawian people (Kureya and Kureya, n.d.). Malawi's reproductive health program is embedded in the Joint Programme of Work for a Health Sector-Wide Approach (2004–2010), implemented by the government with the assistance of development partners (Respond Project 2012). The program's objective was to (1) "establish and deliver an essential health package (including family-planning), to be provided free of charge to all Malawians" and (2) "address the severe shortages of workers in the health sector by improving the retention, training, and deployment of health care staff" (Respond Project 2012, 3).

The individual effects of various policies and programs are difficult to quantify, but between 2004—when efforts to increase access to family-planning began—and 2016—the year of the most recent Demographic and Health Surveys (DHS) with publicly available data—the percentage of currently married women using a modern method of contraception more than doubled from 28 percent (National Statistical Office [Malawi] and ORC Macro 2005) to 58 percent (National Statistical Office and ICF International 2016). Additionally, during the same period, there was a significant expansion from 6 percent to 24 percent in the use of long-acting and permanent methods among currently married women (National Statistical Office and ICF International 2016). Almost three-quarters of modern contraceptive users obtain it at public-sector facilities where services are free. Among injectable users (the most popular method in Malawi), 84 percent go to public-sector facilities and another 9 percent go to Christian Health Association of Malawi (CHAM) facilities (Skiles et al. 2015) (the government contracts with some CHAM facilities to provide the free essential health package in rural areas [SHOPS Project 2012]). Despite progress in expanding access to contraception, however, shortages of both short- and long-term methods remain a problem (Ministry of Health [Malawi] and ICF International 2014), which is likely to affect continuity of use (Respond Project 2012; USAID/Africa Bureau et al. 2012).

Data, Analyses, and Methods

Data

The Malawi Schooling and Adolescent Study (MSAS) is a longitudinal survey that followed 2,649 adolescents (1,337 females) aged 14–17 when first interviewed in 2007. At baseline the sample comprised 1,764 students (875 females) randomly selected from enrollment rosters in randomly selected primary schools in Balaka and Machinga, two rural districts in southern Malawi. The sample also included 885 adolescents (462 females) not enrolled in school, who resided in those schools' catchment villages. Out-of-school adolescents were identified through key informants located at the school or in the randomly selected school catchment villages. Six rounds of data collection were completed, with the last round collected between August 2013 and October 2013. Follow-up rates for the females were: 91 percent in 2008 (round 2), 91 percent in 2009 (round 3), 89 percent in 2010 (round 4), 90 percent in 2011 (round 5), and 89 percent in 2013 (round 6).

The survey collected data on schooling, marital and birth histories, and contraceptive use, among other topics. Audio computer-assisted self-interviews were conducted for sensitive topics, including sexual behavior. In rounds 1–5 of data collection, adolescents were asked about injectable, pill, and condom use, the most commonly used methods of contraception among young women in Malawi reported in the DHS that preceded MSAS in 2004 (National Statistical Office and ICF Macro 2011). In round 6, the MSAS survey section on contraception was greatly expanded. Respondents were asked about ever and current use of modern methods (injectable, implant, pill, intrauterine device [IUD], male condom, female condom, sterilization) and the facility from which current methods were obtained. Those who had previously given birth, were not currently pregnant, and were not currently using contraception were asked reasons for their nonuse.[2]

The Malawi Service Provision Assessment (SPA) was designed to be a census of all formal health care facilities in the country. It was implemented by the Ministry of Health with technical assistance from the DHS program funded by the United States Agency for International Development. The Central Monitoring and Evaluation Division of the Malawi Ministry of Health compiled a master list of all formal-sector health facilities in Malawi ($n = 1,060$) and data were successfully collected on 92 percent of these facilities ($n = 977$). Facilities were of varying types (e.g., hospital, health center, and clinic) and under various managing authorities (e.g., government, Christian Health Association of Malawi, and the private sector). The SPA included a facility inventory, health provider interview, observation, and patient exit interviews. Data collection took place during June–August 2013 and November 2013–February 2014.

These analyses use all rounds of MSAS data and also link round 6 of MSAS data, collected from females in 2013 when respondents were aged 20–23, with SPA data from 2013–2014 to assess the service environment at the time that MSAS respondents reported contraceptive use.

Timing of first use of contraception

We begin with an exploration of the timing of initiation of contraceptive use among the MSAS female sample in relation to first birth and first marriage. This provides useful contextual information for discussion of the way in which facility access and quality may affect the demand for services. Because respondents were not asked the date of first use of contraception, we use the round of data collection (1–6) in which each event was first reported. All rounds of data collection were approximately one year apart (except for a two-year gap between rounds 5 and 6). Ever use of a modern contraceptive method was defined as use of pill or injectable. As condoms may be obtained from sources other than health facilities, condom use was excluded from these analyses (Wang et al. 2012). First birth was considered to have preceded first use of contraception if it was reported in a round prior to or the same round as first report of contraceptive use. Given that injectables are the most common method used in this population (National Statistical Office and ICF Macro 2011), it is less likely that someone first used injectables (rendering her unable to conceive for three months), conceived, and gave birth in the year between two interviews than that she gave birth and then began using contraception, perhaps upon recommendation of a health care provider, sometime in the months after delivery. First marriage was considered to have preceded contraception only if it was reported in a round prior to first report of contraceptive use. We also describe current use of contraception (pill or injectable) within each round, stratified by whether a respondent had ever given birth and her current marital status.

Sociodemographic correlates of contraceptive use

Next, we describe our sample and its contraceptive use at round 6 in 2013 and examine associations between key demographic and socioeconomic characteristics of young women in the MSAS sample (i.e., marital status, parity, and education) and current contraceptive use at round 6 using simple descriptive statistics, bivariate logistic regressions, and a multivariable logistic regression model. Examining the influence of these individual-level characteristics on contraceptive use for a sample of young women sheds light on the development versus service provision debate (for example, by examining the effects of education) in this population. We restrict the sample to the 994 MSAS females interviewed at that round who had ever had sex and were not currently pregnant. Current use of modern contraception was defined as use of injectable, implant, pill, IUD, or sterilization.

Construction of facility-level measures

Facility-level characteristics obtained from the SPA included facility type, managing authority, urban versus rural location, existence of user fees for contraception, and facility readiness indicators. Indicators of facility readiness were constructed based on an analysis that used data from SPA surveys conducted in Tanzania, Kenya, and Ghana (Hutchinson, Do, and Agha 2011). Seventeen measures were created in four domains (termed "structural attributes of quality" by Hutchinson and colleagues): infrastructure and equipment, management, availability of services, and counseling (see appendix). These measures were identical to Hutchinson, Do, and Agha's except where precluded by data limitations and lack of variability among Malawian facilities.[3]

Linking family-planning facilities to MSAS respondents and assessing the impact of services

We linked MSAS respondents to facilities in two ways: (1) based on geographic proximity and (2) based on respondent reports of where they obtained contraception, among those using contraception. First, MSAS respondents were linked to nearby facilities in the SPA using GPS coordinates collected at the time of the interview. The MSAS coordinates were documented at the location of the interview (94 percent of interviews were conducted at the respondent's home). In the case of respondents for whom coordinates were missing, round 5 coordinates were used if the respondents had not reported moving between the two rounds. For the remaining forty-seven women for whom GPS data were missing, we imputed coordinates based on all location data we had for the respondent (e.g., village, traditional authority, maps, and nearby landmarks). All analyses involving linkages using GPS data were also run without these forty-seven cases to check for robustness of results. The SPA dataset included the GPS coordinates of each facility. To explore how the service environment might influence current use of contraception, we identified up to ten facilities within ten kilometers of the respondent (Burgert and Prosnitz 2014). These facilities were identified using the Stata package *geonear*, which computes geodetic distances, and facility characteristics were merged into a joint dataset with MSAS data.

We examined whether both the quantity and quality of facilities within a ten-kilometer radius of respondents were associated with contraceptive use. To assess the impact of service environment quality, we averaged each facility readiness indicator for all such facilities (Stephenson, Beke, and Tshibangu 2008; Wang et al. 2012), creating for each respondent a mean score for each continuous variable or a proportion of facilities with an attribute for each binary variable. If a respondent did not live within ten kilometers of any facility, we assigned the characteristics of the nearest facility. We then averaged these pooled indicators to report the difference between users and nonusers and assessed the associations between the service environment and current use of contraception using logistic regression. As a robustness check, for each indicator we also assigned the highest score among those facilities to each respondent and repeated the analysis. In addition, we estimated a variance-components model to assess whether characteristics of the nearest facility were associated with contraceptive use within our sample (Rabe-Hesketh and Skrondal 2008). We then assessed whether characteristics of the nearest facility were associated with the probability of contraceptive use with bivariate random effects logistic regression models.

Second, we explored factors affecting facility choice among contraceptive users, who reported the facility where they obtained their current method. We received a de-identified list of facilities from the SPA team and matched MSAS responses to SPA facility codes to merge the datasets and acquire the characteristics of facilities that respondents visited for family-planning. Thus, we were able to identify the characteristics of facilities where young women in our sample obtained contraception and assess the individual- and facility-level factors that influenced which facilities they visited. Distance from the respondent to the facility at which she obtained contraception was computed using the Stata package *geodist*. Given the focus on increasing accessibility of family-planning in low prevalence settings, respondents were categorized according to whether they obtained services at the nearest facility providing family-planning.

Among respondents who were currently using contraception, we estimated a variance-components model to evaluate the extent to which facility-level factors explained whether the respondent obtained services at the nearest facility (Rabe-Hesketh and Skrondal 2008), in order to shed light on why some users chose to travel longer distances to obtain contraception. We then assessed the relative contribution of facility- and individual-level characteristics to the probability of going to the nearest facility with bivariate and multivariable multilevel random effects logistic regression models. All analyses were done in Stata 13 (StataCorp 2013).

Results

Timing of first use of contraception

More than half (58 percent) of the female MSAS sample ($n = 1,337$) reported ever using contraception (pill and injectable) by the last round of data collection in 2013. Approximately one-third (34 percent) reported never using contraception by 2013; another 8 percent were not observed in 2013 and had never reported use of contraception prior. The remaining few ($n = 3$) were missing data on contraceptive use at all rounds. For a small group of respondents (approximately 5 percent), the timing of first use of contraception in relation to first birth and first marriage could not be determined either due to reporting contraceptive use and birth and/or marriage at baseline or to missing data.

Among those who reported first use of contraception by 2013 for whom timing in relation to first birth could be ascertained ($n = 683$), the vast majority (87 percent) of young women first used contraception sometime after first birth. Of those who gave birth by the last round in which we observed them ($n = 1,018$), 9 percent used contraception prior to first birth, 58 percent used after first birth, and an additional 33 percent never used and, thus, if they do use contraception, will initiate use after first birth.[4] Similarly, among those who reported first use of contraception by 2013 for whom timing in relation to first marriage could be ascertained ($n = 697$), almost three-quarters (74 percent) first used contraception after marriage, and another 12 percent reported first using contraception in the same round at which they reported first being married. Of those who first married by the last round in which we observed them ($n = 1,065$), 9 percent used contraception prior to first marriage, 49 percent used after, 8 percent used in the same round, and an additional 35 percent never used and, thus, if they do use contraception, will initiate use after marrying.[5] Approximately three-quarters of respondents (73 percent) reported first use of contraception after both birth and marriage, and an additional 10 percent reported first use after birth in the same round as marriage.

Note that, in contrast to pill and injectable use, which the majority of users initiated after first giving birth (87 percent), the majority of condom users ($n = 735$) reported use before (58 percent) or during the same round (13 percent) as their first

FIGURE 1
Current Use of Contraception (Pill and Injectable) among MSAS Females Aged 14–17 in 2007 According to Whether They Had Ever Given Birth and Current Marital Status

birth. Similarly, among those who reported using a condom by 2013 ($n = 765$), more than half (52 percent) used a condom before first marriage, and an additional 14 percent used a condom the same round at which they reported first being married.

Figure 1 displays current use of pill and injectable in each year of data collection, by current marital status and whether a respondent had ever given birth.[6] Those who had ever given birth were more likely to use contraception than those who had not at all rounds of data collection. The largest number of respondents to use contraception were those who were married and had ever given birth.

Sociodemographic correlates of contraceptive use

Of the 1,186 female MSAS respondents interviewed in round 6, 10 percent were pregnant ($n = 120$) and 6 percent had never had sex ($n = 71$). These cases were excluded from all analyses of round 6 data.[7] Additionally, one respondent lived in Mozambique and was therefore excluded. The remaining sample included 994 respondents. More than half of this sample (54 percent) reported current use of modern contraception (injectable, implant, pill, IUD, or sterilization). The vast majority of MSAS respondents using modern contraception were using injectables (80 percent, $n = 422$), followed by implants (17 percent, $n = 89$). Few respondents were currently using pills (3 percent, $n = 14$) or IUDs (2 percent, $n = 9$), and only one reported being sterilized.[8] Among those who were currently using pills or injectables and for whom we knew their contraceptive status in prior rounds ($n = 413$), this was the first reported use for almost half (46 percent). Among those who were not currently using a modern method ($n = 449$), almost two-thirds (65 percent) reported that they had never used modern contraception.

Table 1 shows current use of modern contraception by sample characteristics. Those who were never married (8 percent) or were separated, divorced, or widowed (38 percent) were less likely to use contraception than those who were currently married (62 percent). This effect remained after adjusting for the other covariates listed in the table (AOR, not currently married: 0.36, $p < .001$). As expected, higher parity was positively associated with higher contraceptive use. Belonging to the Yao tribe was associated with being less likely to use contraception in the multivariable model, while belonging to the highest wealth tertile was associated with being more likely to use contraception compared to the lowest tertile.

Access and quality of family-planning services and current contraceptive use

Almost all respondents (90 percent) lived in rural areas (as opposed to urban areas or in a boma [district headquarters]). For only a minority (16 percent), the nearest facility offering family-planning services was located within one kilometer. However, for nearly two-thirds (66 percent) of respondents, the nearest facility was within five kilometers. Another 31 percent lived between five and ten kilometers from the nearest facility, with the remaining few (3 percent) residing ten kilometers or more (see Figure 2). The median number of facilities (up to ten) providing family-planning within a five-kilometer radius of each respondent was only one; the median number of facilities (up to ten) within a ten-kilometer radius was three. There was no significant difference in the mean number of facilities within ten kilometers between users (3.5) and nonusers (3.6).

Descriptive statistics of characteristics of facilities included in the analysis (within ten kilometers of at least one MSAS respondent or the nearest facility to at least one respondent who did not live within ten kilometers of any facility) are listed in the appendix. Descriptive statistics of facility characteristics and their bivariate associations with current contraceptive use are provided in Table 2. The service environment around each respondent is represented two ways for every indicator: the mean value and the maximum value among up to ten facilities within ten kilometers. These indicators were then averaged for users and nonusers to compare the service environments between the two groups. Only one indicator—number of visual aids for demonstrating use of family-planning methods at facility—was significantly

TABLE 1
Current Use of Modern Contraception[a] by Sample Characteristics among MSAS Respondents Age 20–23 in 2013[b]

	N = 994		N = 881[c]			
	%	N	Unadjusted OR	95% CI	Adjusted OR	95% CI
Current age						
20–21	51.6	572	Ref		Ref	
22–23	56.4	422	1.11	0.85, 1.45	0.98	0.73, 1.32
Marital status						
Currently married	62.0	771	Ref		Ref	
Never married[d]	7.9	101	0.28	0.20, 0.41°°°	0.36	0.24, 0.52°°°
Separated, divorced, widow[d]	38.0	121				
Parity						
0	3.2	94	-----		-----	
1	43.5	285	Ref		Ref	
2	65.5	444	2.41	1.77, 3.27°°°	2.16	1.51, 3.09°°°
3+	67.3	171	2.66	1.79, 3.97°°°	2.55	1.57, 4.14°°°
Education						
< Primary	59.7	636	Ref		Ref	
Completed primary	54.5	66	0.91	0.53, 1.55	0.89	0.50, 1.59
Secondary: Partial or completed[e]	39.1	281	0.67	0.49, 0.92°	0.82	0.55, 1.22
Chichewa reading comprehension[f]						
Below the sample mean	55.8	344	Ref		Ref	
Above the sample mean	52.2	647	1.03	0.78, 1.36	1.10	0.80, 1.52
Age at school leaving						
Less than 18	58.8	507	Ref		Ref	
18 and older[g]	53.0	432	0.84	0.64, 1.10	1.13	0.81, 1.59
Currently attending school[g]	10.9	55				
Tribe						
Other	56.0	570	Ref		Ref	
Yao	50.5	424	0.71	0.54, 0.93°	0.70	0.52, 0.93°
Household wealth tertile[h]						
Low	51.6	374	Ref		Ref	
Medium	57.7	312	1.31	0.96, 1.81†	1.21	0.86, 1.70
High	51.9	308	1.43	1.03, 2.00°	1.58	1.09, 2.28°

(continued)

TABLE 1 (CONTINUED)

	N = 994		N = 881[c]			
	%	N	Unadjusted OR	95% CI	Adjusted OR	95% CI
Ever moved since Round 5 (2011)						
No	57.4	544	Ref		Ref	
Yes	49.2	445	0.83	0.63, 1.09	0.87	0.65, 1.17
Urban/boma[i]						
Rural	53.6	893	Ref		Ref	
Urban/boma	53.5	101	1.05	0.67, 1.64	1.01	0.62, 1.64
LR chi² (13)					87.7	○○○

°°°$p < .001$; °°$p < .01$; °$p < .05$; †$p < .10$
NOTE: Unadjusted odds ratios were computed using bivariate logistic regression models. Adjusted odds ratios were computed using a multivariable logistic regression model including all variables listed in Table 1.
a. Defined as use of injectable, implant, pill, IUD, or, sterilization
b. Among round 6 females excluding 120 pregnant respondents, 71 respondents who report never having sex, and 1 respondent who does not live in Malawi.
c. Those with no children were excluded from regressions as there were only three contraceptive users with no children.
d. The never-married and separated, divorced, widowed categories were collapsed in regressions as there were only eight never-married contraceptive users.
e. Only twenty respondents completed secondary.
f. Scored out of 6 (0 if could not read). Sample mean = 3.84.
g. The 18 and older and currently attending school categories were collapsed in regressions as there were only 6 contraceptive users currently attending school (and as the sample was aged 20–23 in 2013, all those currently attending school at round 6 would therefore leave school after age 18).
h. Household wealth tertiles were estimated from principal component analysis of possession of fourteen household items.
i. A boma is a district's headquarters.

associated with current use of contraception ($p < .05$). However, this association was negative, and the difference between users and nonusers was small.[9] Multivariable models including the mean values and the maximum value of facility characteristics shown in Table 2 were not jointly significant (models not shown: likelihood ratio chi square (18) = 22.03, $p = .23$; likelihood ratio chi square (19) = 15.60, $p = .68$, respectively.)[10] Additionally, a variance-components model indicated that only 2 percent of the variation in current use of modern contraception among MSAS respondents was attributable to the characteristics of the nearest facility providing family-planning services (rho = .022; $p < .05$). Bivariate multilevel logistic regression models of the characteristics of the nearest facility (type, managing authority, urban versus rural location, existence of user fees for contraception, and seventeen readiness indicators) did not show any significant association with current contraceptive use (results not shown). We did not

FIGURE 2
Location of Female MSAS Respondents and Health Facilities That Offered Family-Planning Services, 2013

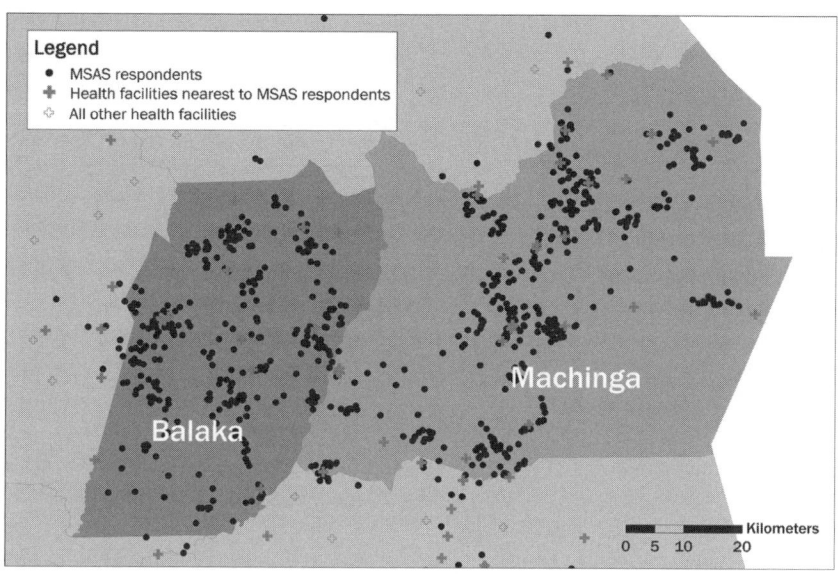

NOTE: The map focuses on Balaka and Machinga, the districts from which our respondents were originally sampled, and where 87 percent of them still lived in 2013. (The remainder not shown here were scattered around the country.)

estimate multivariable models because none of these models was significant, and because of the small value of rho in the variance components model.

Where users obtained their contraceptives

We explored where respondents currently using modern contraception obtained their method of contraception. Of those respondents for whom we had data on where contraception was obtained ($n = 512$), most went to a health center (64 percent), followed by a hospital (19 percent), and clinic or other health facility (10 percent). A small proportion (5 percent) acquired contraception from mobile or outreach services or a health surveillance assistant; for the remaining few (2 percent), we could not identify the facility type.[11] Among respondents who went to facilities listed in the SPA (86 percent, $n = 441$), only 15 percent went to a facility that charged fees for contraception. Interestingly, 15 percent ($n = 64$) of MSAS respondents who visited facilities identified in the SPA reported obtaining contraception at facilities that, according to the SPA, did not provide family-planning services; this type of facility represented 10 percent of the facilities in the SPA to which respondents reported going ($n = 82$ facilities). One facility was a government hospital (one respondent visited); the remaining seven were

TABLE 2
Characteristics of Nearest Facilities (within 10 Kilometers)[a] to Users and Nonusers of Contraception[b]

	Mean value among 10 nearest facilities within 10 km		Maximum value among 10 nearest facilities within 10 km	
	Users[c]	Nonusers[c]	Users[d]	Nonusers[d]
	$N = 533$	$N = 461$	$N = 533$	$N = 461$
Facility characteristics				
Free family-planning provided	66.8%	66.9%	96.4%	95.4%
Urban location	13.2%	11.8%	24.0%	24.1%
Hospital or health center	63.2%	63.2%	93.8%	93.9%
Government facility	60.5%	59.0%	95.3%	94.1%
Infrastructure and equipment				
Physical infrastructure scale (0–5)	3.5	3.6	4.1	4.2
Equipment in family-planning (FP) service area scale (0–7)	4.2	4.3	5.2	5.4†
Management				
Routine facility management meetings	61.2%	62.0%	92.9%	93.5%
System to collect client opinion	45.9%	46.2%	80.7%	83.7%
Quality assurance program	52.3%	54.3%	83.1%	85.2%
Supervisory visit in last six months	88.7%	88.8%	99.8%	99.8%
Stock inventory, organization, and quality scale (0–5)	4.3	4.4	4.8	4.8
Availability of services				
# days FP services provided (0–7)	3.8	3.9	5.1	5.3
# FP methods offered (0–13)	6.7	6.7	6.7	6.7
# FP methods available (0–10)	4.8	4.7	6.5	6.6
Provider available 24 hours	79.0%	81.1%	98.1%	98.3%
# other reproductive health services (0-3)	2.2	2.2	2.8	2.8
Counseling				
FP guidelines	46.1%	47.2%	79.5%	82.6%
# visual aids (0-5)	1.9	2.0°	2.7	2.8°
Has private room	75.7%	75.5%	91.6%	92.0%
Individual client card	39.6%	43.6%†	70.4%	75.1%†
Proportion providers trained in FP	0.3	0.3	0.6	0.6

°°°$p < .001$. °°$p < .01$. °$p < .05$. †$p < .10$.
NOTE: Used bivariate logistic regression to determine significance of associations.
[a]If the respondent did not live within 10 kilometers of any facility, the characteristics of the nearest facility were assigned.
[b]Defined as use of injectable, implant, pill, IUD, or sterilization.
[c]Continuous variables: mean of the respondents' means; binary variables: mean of percent of facilities that met criteria for each respondent.
[d]Continuous variables: mean of the respondents' maximum values; binary variables: percent of respondents that live near at least one facility that met criteria.

CHAM health centers (sixty-three respondents visited). We relied on the respondent report that these facilities did indeed provide family-planning because multiple respondents reported getting contraception from each CHAM facility; in particular, three or more respondents (range: 3–22) stated that they obtained contraception at each of the seven CHAM facilities. Family-planning readiness indicators could not be calculated for these facilities as the data necessary were not collected by the SPA.

Less than half (45 percent, $n = 232$) of respondents obtained contraception at the facility providing family-planning services closest to the place at which they were interviewed in round 6.[12,13] Table 3 displays the characteristics of facilities where respondents got contraception by whether the respondent went to the nearest facility providing family-planning services (among those who went to a facility listed in the SPA). The majority of respondents in both groups went to health centers, although more respondents who did not go to the nearest facility went to hospitals than those who went to the nearest facility (36 percent versus 8 percent). Both groups were equally likely to obtain services from a government facility, but those who did not go to the nearest facility were more likely to pay for contraception. (Additional analyses not shown here demonstrated that those who did not go to the nearest facility lived in areas where a higher proportion of facilities within ten kilometers charged for family-planning). Those who did not go to the nearest facility were far more likely to go to urban facilities. Although those who went to the nearest facility most likely traveled shorter distances than those who did not (depending on the density of facilities near them),[14] it is interesting to note the large percentage of respondents (30 percent) who did not go to the nearest facility and appear to have traveled fifteen kilometers or more to obtain contraception.

Determinants of facility choice

Further exploration demonstrated that migration was a key determinant of whether a respondent went to the nearest facility. Almost half of the round 6 sample (45 percent, $n = 445$) reported having moved (on a visit, temporary, or indefinite basis) at least once since being interviewed at round 5 approximately two years earlier. Of those who moved at least once and were currently using contraception, only 33 percent went to the facility nearest to where they were interviewed in round 6, as compared with 54 percent of those who had never moved since round 5 ($p < .001$). Of those respondents who moved at least once since round 5, 21 percent reported getting contraception at a facility fifteen kilometers or more away (15–445km) versus 9 percent of those who had never moved since round 5 (15–74km). Thus, it appeared that some respondents reported obtaining contraception at a facility in a region from which they had since moved by the time of the round 6 interview. We did not ask when they visited the facility. Still, almost half of the respondents who had not moved did not go to the nearest facility, and some traveled great distances to obtain contraception.

TABLE 3
Characteristics of Facilities Where Respondents Obtained Contraception,[a] by Whether Respondent Went to the Nearest Facility[b]

Facility characteristic	Did not get at nearest facility $n = 209$	Got at nearest facility $n = 232$
Type		° ° °
Hospital	35.9	7.8
Health center	50.7	81.5
Clinic/other	13.4	10.8
Managing authority and fee[c]		° ° °
Government/public: Free	67.9	65.9
CHAM: Free	13.9	25.4
CHAM: Not free	7.2	7.3
Private: Not free	3.3	0.9
NGO: Not free	7.7	0.4
Urban	44.5	7.3°°°
Distance from respondent to facility (km)		° ° °
0–4	32.1	78.0
5–9	30.1	21.1
10–14	7.7	0.9
15 or more[d]	30.1	0

°°°$p < .001$. °°$p < .01$. °$p < .05$. †$p < .10$.
NOTE: Used chi-squared tests to determine significance of associations.
[a]Defined as use of injectable, implant, pill, IUD, or, sterilization.
[b]Sample includes only current users of contraception who obtained contraception at a facility listed in the SPA, as we do not know facility characteristics for those who did not go to a facility listed in the SPA. Facility characteristics are reported per respondent (more than one respondent may report going to a facility and the sample size is the number of respondents).
[c]Tested using two-sided Fisher's exact test due to expected frequency < 5 in at least one of the cells. Among respondents who went to government facilities, 2 percent went to facilities that reported charging fees for contraception in the SPA. The SPA reports whether a fee is charged for contraception at all facilities, regardless of whether it reports family-planning services are available there.
[d]Maximum distance is 445 km between facility where respondent reported obtaining contraception and GPS coordinates recorded at interview.

We also explored other individual- and facility-level correlates of visiting the nearest facility to obtain contraception. A variance-components model indicated that more than half (56 percent) of the variation in visiting the nearest facility among MSAS respondents was attributable to characteristics of the nearest facility (rho = .563; $p < .001$). The results were essentially the same (52 percent, rho = .516, $p < .001$) among those who had never moved since round 5.

Bivariate multilevel logistic regression models were estimated with each of the seventeen facility readiness indicators to determine if they were associated with whether a respondent obtained contraception at the nearest facility.[15] Only four were significant: number of days per week that family-planning services were provided, availability of provider 24 hours a day, supervisory visit within the last six months, and number of other reproductive health services offered (results not shown).[16] After these variables were added to a model with other facility characteristics (facility type, fee, and urban/rural location), only the "number of other reproductive health services offered" remained significant (results not shown). Further analysis revealed that these four characteristics were strongly associated with facility type and fee (among the facilities nearest to this sample of respondents, results not shown). Facilities that charged fees had family-planning services available more days per week than those that did not charge a fee. Health centers and hospitals almost universally had a provider available 24 hours a day, and all health centers had a supervisory visit within the past six months. Free facilities were more likely to offer a greater number of other reproductive health services than facilities that charged fees. All hospitals and health centers offered two to three other reproductive health services (with approximately 80 percent offering three other services). In contrast, 84 percent of clinics offered either no other service or just one. Due to the strong association between other services and facility type, "other services" was excluded from the final model. Managing authority was not included in the model with fees as there was little variation within categories. Among facilities nearest to respondents, virtually 100 percent of government facilities were free, whereas nearly all private and NGO facilities charged fees for contraception. Half of CHAM facilities were free and the other half charged for contraception.

In a multivariable multilevel logistic regression model estimating only the associations between individual-level characteristics and facility choice (with women nested within facilities), only three variables were significantly predictive of not going to the nearest facility for contraception: ever having moved since round 5, distance of five kilometers or more to nearest health facility, and being in the middle or highest wealth tertile as compared to the poorest (results not shown). The final multivariable multilevel logistic regression model (Table 4) included individual-level characteristics and facility characteristics with no readiness indicators. The individual-level characteristics significant in the model described above (with no facility-level characteristics) remained significantly associated with going to the nearest facility for contraception (with the exception of being in the highest wealth tertile). Those who moved since round 5 were less likely to go to the nearest facility (AOR: 0.37, $p < .001$). Those who lived five kilometers or farther from the nearest health facility were less likely to go to it than those who lived fewer than five kilometers away (AOR: 0.19, $p < .001$). In other words, if a woman did not live within five kilometers of a facility, she was less inclined to go to the closest facility. Respondents in the middle wealth tertile were less likely to go to the nearest facility than the poorest tertile (AOR: 0.47, $p < .05$). Respondents who lived nearest to facilities that provided free contraception were more likely to go to those facilities (AOR: 8.98, $p < .001$). We tested a model including an interaction term for distance and cost; the interaction was not significant and therefore was not included in the final model. Those who lived near clinics or other facility types were less likely to go to them than those who lived near health centers (AOR: 0.23, $p < .001$).

TABLE 4
Multivariable Multilevel Logistic Regression Model of Whether Current Users of Contraception[a] Went to the Nearest Facility to Obtain It

N = 435[b]	AOR	95% CI
Current age		
20–21	Ref	
22–23	1.18	0.70, 1.98
Marital status		
Currently married	Ref	
Not currently married	0.83	0.35, 1.96
Parity		
1	Ref	
2	1.35	0.70, 2.63
3+	1.35	0.58, 3.13
Education		
< Primary	Ref	
Completed primary or greater	1.31	0.68, 2.50
Chichewa reading comprehension[c]		
Below the sample mean	Ref	
Above the sample mean	1.44	0.82, 2.51
Tribe		
Other	Ref	
Yao	1.47	0.85, 2.53
Household wealth tertile[d]		
Low	Ref	
Medium	0.47	0.25, 0.86°
High	0.55	0.29, 1.04
Ever moved since round 5 (2011)		
No	Ref	
Yes	0.37	0.22, 0.64°°°
Distance to nearest health facility offering family-planning services (km)		
<5	Ref	
5+	0.19	0.10, 0.36°°°
Nearest facility type		
Health center	Ref	
Hospital	2.74	0.73, 10.28
Clinic/other	0.23	0.10, 0.56°°°
Cost of family-planning at nearest facility		
Not free	Ref	
Free	8.98	3.55, 22.71°°°
Nearest facility location		
Rural	Ref	
Urban	0.33	0.10, 1.15
Wald Chi² (15)	74.36	°°°
Rho	0.13	°°

°°°$p < .001$. °°$p < .01$. °$p < .05$. †$p < .10$.
[a]Defined as use of injectable, implant, pill, IUD, or, sterilization.
[b]Three respondents with no children were excluded from this model. The sixty-four respondents who obtained contraception at a facility that did not provide family-planning according to the SPA were excluded because quality indicators on the nearest facility were missing for respondents who obtained contraception at facilities nearer to them than the "nearest" facility according to the SPA.
[c]Scored out of 6 (0 if could not read). Sample mean = 3.84.
[d]Household wealth tertiles were estimated from principal component analysis of possession of fourteen household items.

Reasons why those who had previously used contraception were not currently using it

Finally, we explored reasons for nonuse among those who had used contraception (injectable, implant, pill, IUD, or sterilization) in the past, were not currently using, and yet reported wanting to space or limit births ($n = 98$)[17]; the largest proportion indicated they were not having sex (35 percent). The next most common reasons cited for nonuse were breastfeeding (27 percent), not being married (18 percent), not having menstruated since last birth (13 percent), side effects/health concerns (12 percent), and opposition to contraceptive use (8 percent). No respondent stated that she was unaware of a place to obtain contraception or that access, distance, cost, not being able to get her preferred method, or stockout was a problem.

Robustness considerations

To investigate robustness of results that describe the relationship between access and current contraceptive use, we estimated the models using alternative measures of facilities and distances.

- A categorical number of facilities (rather than continuous) in ten-kilometer radius was not significantly associated with current contraceptive use.
- Bivariate associations between current contraceptive use and characteristics of facilities within a five-kilometer radius were not meaningfully different from those within a ten-kilometer radius (as shown in Table 2); no bivariate associations between current contraceptive use and characteristics of facilities within a ten-kilometer radius were significant in an analysis limited to rural respondents.
- Distance to the nearest health facility offering family-planning services was not significantly associated with current use of contraception.

We also assessed the robustness of our results that describe whether respondents went to the nearest facility (Table 4). These did not meaningfully change when we restricted the sample to respondents who had not migrated since round 5, nor did they change when we restricted the analysis to respondents living in rural areas.

Finally, all analyses involving data linked by GPS coordinates were repeated without the cases for which coordinate data were imputed. There were no meaningful differences in the results.

Discussion

In a setting where family-planning programs have expanded such that there appears to be little variability in access, at least as we are able to measure it, demographic characteristics appear to be the primary factor affecting contraceptive use among young women. Both the longitudinal and the round 6 data from MSAS show that current users of contraception were much more likely than nonusers to be married and to have given birth. Very few young women began to use contraception prior to first birth and first marriage. This echoes findings from the DHS that most women in Malawi begin using contraception after their first birth (National Statistical Office and ICF Macro 2011). Lack of use may be due to stigma surrounding contraceptive use among unmarried women for whom use is a tacit acknowledgement of being sexually active (Williamson et al. 2009; Wood and Jewkes 2006). In addition, those who wish to use contraception but have not had children may be discouraged from using or even denied it (Prata, Weidert, and Sreenivas 2013). Such provider "bias" might partially be addressed by sensitizing and expanding programs; however, altering the service environment may have little effect in the absence of wider normative change. Moreover, whether unmarried or married, adolescents and young women may not want to delay or prevent their first birth due to pressure to demonstrate their fecundity (Hindin and Fatusi 2009; Preston-Whyte et al. 1990; Williamson et al. 2009; Wood and Jewkes 2006). Jewkes et al. (2001) suggest that although the majority of adolescent pregnancies are unplanned, they are not necessarily unwanted. It follows that expanding access to family-planning services, high quality or otherwise, may not translate into substantially lower fertility in these populations.

The presence and characteristics of nearby facilities with contraception available did not appear to affect use among young women in our sample. Approximately 2 percent of the variation in current use of modern contraception among MSAS respondents was attributable to the facility-level characteristics of the nearest facility providing family-planning services. Although this variance components model only accounted for the nearest facility, this finding was consistent with our null findings for the effect of the broader service environment on use of contraception. In contrast, Skiles et al. (2015) found that distance to injectable services sites based on kernel density mapping affected injectable use among rural Malawian women aged 15–49. Those in the two highest quintiles of access were more likely to use contraception than those who lived in the lowest quintile of access. Our results may differ because we focus on use among those at the beginning of their reproductive years, for whom we are predicting first or early use of contraception. This population might have different needs and expectations and face different constraints than those who are older. As a higher percentage of our sample resided in rural areas than in the nation as a whole, we may not have been able to capture the variation in access that exists throughout the country. Still, our results are consistent with other research indicating that geographic access is not a primary barrier to use or cause of unmet need (Bongaarts and Bruce 1995; Choi, Fabic, and Adetunji 2016). In light of the debate on which factors contribute to declines in TFR in high-fertility countries, our results suggest that, whereas access to sites offering contraception is a prerequisite for use by young women, it is by no means sufficient, especially if access exceeds a threshold that most young women do not consider to be problematic. Of note in our sample, among prior users who were not currently using contraception but wanted to space or limit births, no one stated she was unaware of a facility from which to obtain contraception or that access, distance, cost, not being able to get her preferred method, or stockout was the reason that she was not currently using contraception. Additionally, at least among this age-restricted, rural sample, education does not explain variation in current use; rather, marital status and parity are the critical factors that appear to affect demand. Our findings add nuance to the debate by suggesting that the relative importance of family-planning services vs. development may depend on the motivation for using contraception and the level of access already present. Given the expansion of services in Malawi and the fact that young women primarily aim to space births rather than delay a first birth or limit childbearing, the debate regarding the relative contributions of expanding access versus development may have less salience.

In contrast to the models estimating the likelihood of contraceptive use, facility-level characteristics were key determinants of where those who decided to use contraception obtained it. Although facility-level characteristics of the nearest service site explained only a small percentage of the variation in contraceptive use, they accounted for more than half of the variation in visiting the nearest facility among MSAS respondents. The only individual-level characteristics associated with going to the nearest facility were migration, distance to facility, and being in the middle wealth tertile. Young women were more likely to go to facilities that provided free contraception, were health centers as compared to clinics, and were within five kilometers of where they lived. One explanation for our finding that a respondent was less likely to go to the nearest facility if it was not within five kilometers is that she went to the facility that was more convenient in terms of transport even if it was farther away. It is of note that if a respondent did not prefer to go to the nearest facility, but was motivated to use contraception, she sometimes traveled great distances to do so. This fact highlights that, in our sample of young women, proximate geographic access may be less important than other factors in influencing usage.

This study has some important limitations. First, MSAS data were not designed for the purposes to which we have put them. The focus of MSAS was on education, transitions to adulthood, and sexually transmitted diseases. Thus, minimal data on contraception were collected prior to round 6, and we did not ask whether respondents used implants in rounds 1–5, which we found in round 6 to be the second most common method in our sample. Additionally, as our measure of contraception in analyses using all rounds of data included only pill and injectable (excluding condom use), we are providing a somewhat distorted picture of the timing of contraceptive initiation relative to marriage and first birth. Second, gaps remain in data collected at round 6, in particular, an absence of questions on exact timing of first contraceptive use as well as continuity of use. In addition, we did not collect qualitative data, which would have enabled us to investigate decision-making around contraceptive use. Also, the SPA's lack of client observation and interview data for all facilities prevented creation of more comprehensive quality of care measures. Finally, it is somewhat concerning that 10 percent ($n = 8$) of the facilities identified in the SPA where respondents reported getting contraception did not provide family-planning services according to the SPA. Although some of these facilities may have stopped providing contraception between the last time a respondent obtained contraception and the time of the SPA interview, it is improbable that this is the case for all of them. Thus, there is likely some misclassification in the SPA regarding which facilities provided family-planning services due to reporting errors by facility staff, data entry error, or deliberate misreporting to enable skipping large sections of the questionnaire (akin to the age-heaping seen in the DHS to limit eligibility for certain questions [Borkotoky and Unisa 2014; Pullum 2006]).

This analysis underscores the problems that arise from matching datasets based on GPS coordinates, even when the datasets are collected contemporaneously. Some respondents had clearly moved since the time they had last obtained contraception. This resulted in misclassification when determining whether they went to the nearest facility to obtain contraception: where they obtained contraception may have been the nearest facility to them at their previous location, but now did not appear to be the nearest facility because they had moved. The amount of misclassification is difficult to quantify given the highly mobile sample. Additionally, shorter distance as the crow flies does not equate to ease of access for a respondent; this may explain, in part, why a minority of respondents went to the facility nearest them.

The expansion of facilities providing family-planning has led to relatively high prevalence of modern contraceptive use, but fertility is not as low as expected given this prevalence. Many of the young women in our sample may use contraception sporadically (Furnas 2016). Indeed, Furnas (2016) found that many young people's contraceptive behavior in Malawi apparently changes in response to transitions between one partner and another, and changes in preferences within relationships, so that unmet need for contraception may be more dynamic than previously assumed. Previous research has also found that some women purposely delay obtaining their next injectable due to the desire to menstruate (Jewkes et al. 2001; Wood and Jewkes 2006). Furthermore, even though respondents do not cite access as problematic, for the vast majority of users who rely on injectables, having to return to a health facility every few months may be a deterrent to consistent use, particularly in poor rural areas where many women travel long distances to obtain contraception. In other words, our findings are not inconsistent with the results of Skiles et al. (2015) discussed above; access may matter less in our sample, both because we are more likely to be modeling initiation of use rather than continuity of use, and because uninterrupted usage may be a more pressing need for those who are older and more likely to be limiting, rather than spacing, births. The high levels of migration in this population may interrupt usage if young women who have temporarily moved prefer to obtain contraception at a facility that they have visited before or if they delay seeking care at a new facility after a permanent move. In addition, views about the desirability of lower fertility or longer birth intervals may not be that deeply held in the face of the exigencies of daily life. Additional research is warranted on the continuity of use among adolescents and young women in settings such as Malawi.

The contraceptive choices of young women in Malawi and sub-Saharan Africa will be instrumental in defining future population growth. Although many studies have examined the influence of facility access and quality on contraceptive use among all women of reproductive age, the factors influencing use among young women, who face pressures to conceive, likely differ. There has been progress in expanding access to family-planning in Malawi such that almost all respondents in our mostly rural sample live within ten kilometers of a health facility. Our findings indicate that whereas improving access and quality of family-planning services undoubtedly matters for older women, at least in the short term, this may not markedly increase use of contraception among young women without broader shifts in cultural norms regarding childbearing in the early years of marriage.

Appendix

TABLE A1
Family-Planning (FP) Facility Readiness Indicators

Indicator	Definition
Infrastructure and equipment	
Physical infrastructure	Number of amenities available at facility: electricity, water, working toilet, working telephone, waiting area for clients (out of five)
Equipment in FP service area	Number of following items present: exam bed or couch, functioning light, soap, water for handwashing, latex gloves, disinfectant, sharps container (out of seven)
Management	
Routine facility management meetings	Whether the facility has routine management meetings at least once per month
System to collect client opinion	Whether there is a system to obtain clients' opinions regarding services
Quality assurance program	Whether the facility has a routine program for quality monitoring
Supervision	Whether the last supervisory visit to the facility was in the last six months
Stock inventory, organization, and quality[a]	Number of following conditions present at facility: inventory for contraceptive supplies updated daily; stock organized by expiry date; contraceptives protected from water, sun, and pests (out of five)
Availability of services	
Number of days services provided	Number of days per week that FP services are provided
Availability of provider	Whether a trained provider is available 24 hours a day at the facility or officially on call

(continued)

TABLE A1 (CONTINUED)

Indicator	Definition
FP methods offered	Number of methods offered (provide, prescribe, counsel, refer): combined oral pill, progestin-only pill, combined injectable, progestin-only injectable, male condoms, female condoms, IUD, implant, emergency contraceptive pill, cycle beads for standard days method, vasectomy, tubal ligation, other methods (e.g., spermicide or diaphragm) (out of thirteen)
FP methods available	Number of methods available (observed at least one valid): combined oral pill, progestin-only pill, combined injectable, progestin-only injectable, male condoms, female condoms, IUD, implant, emergency contraceptive pill, cycle beads for standard days method (out of ten)
Other reproductive health services offered	Number of RH services besides FP offered: STI services, antenatal care, and delivery (out of three)
Counseling	
Guidelines	Whether any FP guidelines are observed
Visual aids	Number of visual aids for demonstrating use of FP methods at facility: Sample of FP method, other FP-specific visual aids (e.g., flip charts, leaflets), pelvic model for IUCD, model for showing male condom use, model for showing female condom use (out of five)
Privacy	Whether facility has private room for FP services
Individual client card	Whether there is an individual client card/record for FP
Providers trained in FP[a]	Proportion of providers at facility received FP in-service training in past 24 months

*Note: Hutchinson, Do, and Agha (2011) also include the number of years of experience providers have in providing FP services, but the Malawi SPA did not have this information.
[a]If a facility were missing information on this indicator, it was assigned the mean value for either urban or rural facilities, depending on its location.

TABLE A2
Characteristics of Facilities Within Ten Kilometers of at Least One MSAS Respondent[a]
(Mean Unless Otherwise Indicated)

	Government/ Public	CHAM/ Other mission	Private	NGO	Total
N	120	36	130	26	312
Facility characteristics					
Free family-planning provided	97.5%	55.6%	16.9%	19.2%	52.6%
Urban location	31.7%	27.8%	80.0%	84.6%	55.8%
Facility type					
Hospital	17.5%	13.9%	10.0%	0.0%	12.5%
Health center	60.0%	66.7%	1.5%	7.7%	32.1%
Clinic/other	22.5%	19.4%	88.5%	92.3%	55.4%
Infrastructure and equipment					
Physical infrastructure scale (0–5)	3.6	3.7	4.1	4.4	3.9
Equipment in FP service area scale (0–7)	4.4	4.8	5.4	5.9	5.0
Management					
Routine facility management meetings	58.3%	63.9%	33.8%	73.1%	50.0%
System to collect client opinion	53.3%	38.9%	34.6%	96.2%	47.4%
Quality assurance program	56.7%	52.8%	36.9%	73.1%	49.4%
Supervisory visit in last six months	90.0%	91.7%	60.0%	88.5%	77.6%
Stock inventory, organization, and quality scale (0–5)	4.5	4.6	4.2	4.8	4.4
Availability of services					
# days FP services provided (0–7)	3.7	3.7	5.5	5.7	4.6
# FP methods offered (0–13)	7.5	6.9	6.2	9.4	7.0
# FP methods available (0–10)	5.3	4.8	3.7	6.7	4.7
Provider available 24 hours	90.8%	86.1%	46.9%	7.7%	65.1%
# other reproductive health services (0–3)	2.5	2.8	1.4	1.2	2.0
Counseling					
FP guidelines	50.0%	50.0%	48.5%	57.7%	50.0%
# visual aids (0–5)	2.1	2.0	1.5	2.9	1.9
Has private room	81.7%	69.4%	86.2%	84.6%	82.4%
Individual client card	36.7%	41.7%	35.4%	38.5%	36.9%
Proportion providers trained in FP	0.3	0.2	0.3	0.6	0.3

[a] Or nearest facility to at least one MSAS respondent who did not live within ten kilometers of any facility.

Notes

1. At the time this article was written, the DHS had released the 2015–16 Key Indicators Report cited here, but not the complete 2015–16 Malawi Demographic and Health Survey Data and Results.
2. In rounds 1–4, female respondents were not asked directly if they were currently pregnant. In round 5, female respondents were asked if they were currently pregnant, but were not excluded from questions on current use of contraception if they were. In round 6, female respondents were asked if they were currently pregnant, and those responding "yes" were excluded from questions about current contraceptive use.
3. Hutchinson and colleagues (2011) also created quality indicators in two domains (termed "process attributes of quality"), interpersonal and technical, that we were unable to generate because client observation and exit data, from which these indicators could be constructed, were available for only 47% of facilities (n = 380) that provided family-planning services (n = 810). We therefore constructed only readiness indicators.
4. An additional 227 women in the sample had not given birth and not initiated use and thus are censored on both variables.
5. An additional 194 women in the sample had not married and not initiated use and thus are censored on both variables.
6. The sample includes those who were pregnant because respondents were not asked directly at all rounds whether they were pregnant. Also, this sample does not exclude those who reported never having had sex due to high levels of inconsistency in reporting about sexual behavior among adolescents in sub-Saharan Africa (Beguy et al. 2009; Mensch et al. 2014; Palen et al. 2008; Soler-Hampejsek et al. 2013). One analysis found that 55.1 percent of females in this sample reported sex inconsistently at baseline (Soler-Hampejsek et al. 2013). In addition, some respondents who report never having sex report contraceptive use. For example, in round 3, 9.5 percent of those who state they had never had sex report ever using pill or injectable, and 3.6 percent were currently using pill or injectable. (Reports of condom use and sex are even more inconsistent than pill and injectable use and sex: 15 percent of those who report they never had sex in round 3 report ever using a condom.)
7. In contrast to the earlier analysis, here we exclude those who did not have sex; even though a small number of those who had never had sex reported ever using a modern method (injectable, implant, pill, IUD, or sterilization) (5.7 percent, n = 4), none of them reported current use at the time of round 6.
8. An additional thirty-five respondents reported using condoms (and were not using any of the modern methods described here).
9. We did not include a measure of provision of youth-friendly services in our results because the SPA measures are based on self-reports from staff and, thus, are not objective indicators of the degree to which a family-planning provider is welcoming to young women, particularly those who are not married or of low parity. Moreover, upon our examination of what is included in the SPA (the proportion of staff at a facility who report providing "youth-friendly or adolescent-friendly" services), we found no evidence that the existence of youth-friendly services had an effect on contraceptive use (when constructed in the same manner as the variables in Table 2, using the mean and maximum value within ten kilometers).
10. "Government facility" was excluded from both models as government facilities were virtually all free and other managing authorities (namely, private and NGO) were unlikely to have hospitals or health centers (see appendix). "Number of family-planning methods offered" was excluded as it was represented by "number of family-planning methods available." Next, as some of the remaining variables were correlated, we checked the variance inflation factors to test for multicollinearity. This led to the exclusion of the "other reproductive health services available" in the mean values model, as the variance inflation factor was greater than ten.
11. Of condom users in our sample not using any other method of contraception who reported where they obtained condoms (n = 34) in round 6, almost three-quarters (70.6 percent) report they procured them at a health facility (with the largest group of those going to a health center). The remainder got it at either a shop or youth center. However, the vast majority of this sample was currently married (73.5 percent) and had ever given birth (79.4 percent). Thus, this may not reflect where unmarried, nulliparous youth obtain condoms.

12. The sixty-four respondents who reported obtaining contraception at facilities that did not provide family-planning according to the SPA were categorized as going to the nearest facility if the facility that they reported visiting was closer to them than the nearest facility providing family-planning according to the SPA. They were categorized as not going to the nearest facility if the facility that they visited was farther away than the nearest facility providing family-planning according to the SPA.

13. A slightly smaller percentage (26.5 percent) of condom users ($n = 34$) reported getting condoms at the nearest facility providing family-planning.

14. For example, a respondent in an urban area might travel one kilometer to the second-nearest facility to her, but a respondent in a rural area might travel five kilometers to the nearest facility to her.

15. The sixty-four respondents who obtained contraception at a facility that did not provide family-planning according to the SPA were excluded from these multilevel models (including Table 4) because quality indicators on the nearest facility were missing for respondents who obtained contraception at facilities nearer to them than the "nearest" facility according to the SPA.

16. The existence of youth-friendly services (the proportion of staff at a facility who report providing "youth-friendly or adolescent-friendly" services) had no effect on whether a respondent went to the nearest facility.

17. Nine respondents who had never given birth were excluded from these data because they were not asked about their spacing/limiting preferences or reasons for nonuse given inclusion criteria for this question.

References

African Union Commission. 2006. *Plan of action on sexual and reproductive health and rights (Maputo plan of action)*. Addis Ababa: African Union. Available from http://pages.au.int/sites/default/files/static/carmma/MPoA.pdf?q=pages/sites/default/files/static/carmma/MPoA.pdf.

Ainsworth, Martha, Kathleen Beegle, and Andrew Nyamete. 1996. The impact of female schooling on fertility and contraceptive use: A study of fourteen sub-Saharan countries. *The World Bank Economic Review* 10 (1): 85–122.

Akin, John S., and Jeffrey J. Rous. 1997. Effect of provider characteristics on choice of contraceptive provider: A two-equation full-information maximum-likelihood estimation. *Demography* 34 (4): 513–23.

Alkema, Leontine, Vladimira Kantorova, Clare Menozzi, and Ann Biddlecom. 2013. National, regional, and global rates and trends in contraceptive prevalence and unmet need for family-planning between 1990 and 2015: A systematic and comprehensive analysis. *The Lancet* 381 (9878): 1642–52.

Beguy, Donatien, Caroline W. Kabiru, Evangeline N. Nderu, and Moses W. Ngware. 2009. Inconsistencies in self-reporting of sexual activity among young people in Nairobi, Kenya. *Journal of Adolescent Health* 45 (6): 595–601.

Behrman, Julia Andrea. 2015. Does schooling affect women's desired fertility? Evidence from Malawi, Uganda, and Ethiopia. *Demography* 52 (3): 787–809.

Biddlecom, Ann E., Alister Munthali, Susheela Singh, and Vanessa Woog. 2007. Adolescents' views of and preferences for sexual and reproductive health services in Burkina Faso, Ghana, Malawi and Uganda. *African Journal of Reproductive Health* 11 (3): 99–100.

Bongaarts, John. 1994. Population policy options in the developing world. *Science* 263 (5148): 771–76.

Bongaarts, John, and Judith Bruce. 1995. The causes of unmet need for contraception and the social content of services. *Studies in Family-Planning* 26 (2): 57–75.

Bongaarts, John, and John Casterline. 2012. Fertility transition: Is sub-Saharan Africa different? *Population and Development Review* 38 (Supplement s1): 153–68.

Borkotoky, Kakoli, and Sayeed Unisa. 2014. Indicators to examine quality of large scale survey data: An example through district level household and facility survey. *PLoS ONE* 9 (3): e90113.

Burgert, Clara R., and Debra Prosnitz. 2014. *Linking DHS household and spa facility surveys: Data considerations and geospatial methods*. DHS Spatial Analysis Reports 10. Rockville, MD: ICF International.

Casterline, John B. 2001. The pace of fertility transition: National patterns in the second half of the twentieth century. *Population and Development Review* 27 (Supplement: Global Fertility Transition): 17–52.

Casterline, John B., and Steven W. Sinding. 2000. Unmet need for family-planning in developing countries and implications for population policy. *Population and Development Review* 26 (4): 691–723.

Chimbiri, Agnes M. 2006. Development, family change, and community empowerment in Malawi. In *African families at the turn of the 21st century*, eds. Yaw Oheneba-Sakyi and Baffour K. Takyi, 227–43. London: Praeger.

Choi, Yoonjoung, Madeleine Short Fabic, and Jacob Adetunji. 2016. Measuring access to family-planning: Conceptual frameworks and DHS data. *Studies in Family-Planning* 47 (2): 145–61.

Cooper, Diane, Jane Harries, Landon Myer, Phyllis Orner, and Hillary Bracken. 2007. "Life is still going on": Reproductive intentions among HIV-positive women and men in South Africa. *Social Science & Medicine* 65 (2): 274–83.

Defo, Barthelemy Kuate. 2011. The importance for the MDG4 and MDG5 of addressing reproductive health issues during the second decade of life: Review and analysis from times series data of 51 African countries. *African Journal of Reproductive Health* 15 (2): 9–30.

Demeny, P. 1992. Policies seeking a reduction of high fertility: A case for the demand side. *Population and Development Review* 18 (2): 321–32.

Do, Mai, and Nami Kurimoto. 2012. Women's empowerment and choice of contraceptive methods in selected African countries. *International Perspectives on Sexual and Reproductive Health* 38 (1): 23–33.

Dyer, S. J. 2007. The value of children in African countries: Insights from studies on infertility. *Journal of Psychosomatic Obstetrics and Gynecology* 28 (2): 69–77.

Furnas, H. E. 2016. Capturing complexities of relationship-level family-planning trajectories in Malawi. *Studies in Family-Planning* 47 (3): 205–21.

Garenne, Michel, Stephen Tollman, and Kathleen Kahn. 2000. Premarital fertility in rural South Africa: A challenge to existing population policy. *Studies in Family-Planning* 31 (1): 47–54.

Hindin, Michelle J., and Adesegun O. Fatusi. 2009. Adolescent sexual and reproductive health in developing countries: An overview of trends and interventions. *International Perspectives on Sexual and Reproductive Health* 35 (2): 58–62.

Hutchinson, Paul L., Mai Do, and Sohail Agha. 2011. Measuring client satisfaction and the quality of family-planning services: A comparative analysis of public and private health facilities in Tanzania, Kenya and Ghana. *BMC Health Services Research* 11 (203). doi:10.1186/1472-6963-11-203.

Jain, Anrudh. 1989. Fertility reduction and the quality of family-planning services. *Studies in Family-Planning* 20 (1): 1–16.

Jain, Anrudh K., and John A. Ross. 2012. Fertility differences among developing countries: Are they still related to family-planning program efforts and social settings? *International Perspectives on Sexual and Reproductive Health* 38 (1): 15–22.

Jain, Aparna, John Ross, Erin McGinn, and Jay Gribble. 2014. *Inconsistencies in the total fertility rate and contraceptive prevalence rate in Malawi*. Washington, DC: Futures Group, Health Policy Project.

Jewkes, R., C. Vundule, F. Maforah, and E. Jordaan. 2001. Relationship dynamics and teenage pregnancy in South Africa. *Social Science & Medicine* 52 (5): 733–44.

Johnson, Kiersten, Noureddine Abderrahim, and Shea O. Rutstein. 2011. *Changes in the direct and indirect determinants of fertility in sub-Saharan Africa*. DHS Analytical Studies 23. Calverton, MD: Macro International Inc.

Khan, Shane, and Vinod Mishra. 2008. *Youth reproductive and sexual health*. DHS Comparative Reports 19. Calverton, MD: Macro International Inc.

Koenig, Michael A., Mian Bazle Hossain, and Maxine Whittaker. 1997. The influence of quality of care upon contraceptive use in rural Bangladesh. *Studies in Family-Planning* 28 (4): 278–89.

Kureya, Tendayi, and Cynthia Kureya. n.d. *Success story - towards universal access to comprehensive sexual & reproductive health services*. Pretoria: Southern Africa HIV and AIDS Information Dissemination Service. Available from http://www.safaids.net/files/MALAWI_Implementation_of_MPoA_SRHR_services_final.pdf.

Lloyd, Cynthia B. 2005. *Growing up global: The changing transitions to adulthood in developing countries*. Washington, DC: National Academy of Sciences.

Lutz, Wolfgang. 2014. A population policy rationale for the twenty-first century. *Population and Development Review* 40 (3): 527–44.

Magnani, Robert J., David R. Hotchkiss, Curtis S. Florence, and Leigh Anne Shafer. 1999. The impact of the family-planning supply environment on contraceptive intentions and use in Morocco. *Studies in Family-Planning* 30 (2): 120–32.

Malawi National Statistical Office. 2008. *2008 population and housing census main report*. Zomba: Malawi National Statistical Office. Available from http://www.nsomalawi.mw/images/stories/data_on_line/demography/census_2008/Main%20Report/Census%20Main%20Report.pdf.

Mensch, Barbara, Mary Arends-Kuenning, and Anrudh Jain. 1996. The impact of the quality of family-planning services on contraceptive use in Peru. *Studies in Family-Planning* 27 (2): 59–75.

Mensch, Barbara S, Erica Soler-Hampejsek, Christine Kelly, Paul Hewett, and Monica J. Grant. 2014. Challenges in measuring the sequencing of life events among adolescents in Malawi: A cautionary note. *Demography* 51 (1): 277–85.

Ministry of Development Planning and Cooperation. 2010. *Rapid: Population and development in Malawi*. Lilongwe, Malawi: Population Unit, Ministry of Development Planning and Cooperation.

Ministry of Health [Malawi] and ICF International. 2014. *Malawi service provision assessment (MSPA) 2013–14*. Lilongwe, Malawi, and Rockville, MD: Ministry of Health and ICF International.

National Statistical Office [Malawi] and ICF International. 2016. *Malawi Demographic and Health Survey 2015-16: Key Indicators Report*. Zomba, Malawi, and Rockville, MD: NSO and ICF International.

National Statistical Office and ICF Macro. 2011. *Malawi demographic and health survey 2010*. Zomba, Malawi, and Calverton, MD: NSO and ICF Macro.

National Statistical Office and Macro International Inc. 1994. *Malawi demographic and health survey 1992*. Zomba, Malawi, and Calverton, MD: NSO and Macro International.

National Statistical Office [Malawi] and ORC Macro. 2001. *Malawi demographic and health survey 2000*. Zomba, Malawi, and Calverton, MD: National Statistical Office and ORC Macro.

National Statistical Office [Malawi] and ORC Macro. 2005. *Malawi demographic and health survey 2004*. Calverton, MD: NSO and ORC Macro.

Palen, Lori-Ann, Edward A. Smith, Linda L. Caldwell, Alan J. Flisher, Lisa Wegner, and Tania Vergnani. 2008. Inconsistent reports of sexual intercourse among South African high school students. *Journal of Adolescent Health* 42 (3): 221–27.

Population Reference Bureau. 2014. *2014 world population data sheet*. Washington, DC: Population Reference Bureau. Available from http://www.prb.org/pdf14/2014-world-population-data-sheet_eng.pdf.

Prata, Ndola, Karen Weidert, and Amita Sreenivas. 2013. Meeting the need: Youth and family-planning in sub-Saharan Africa. *Contraception* 88 (1): 83–90.

Preston-Whyte, Eleanor, Maria Zondi, Gladys Mavundla, and Hilda Gumede. 1990. Teenage pregnancy, whose problem? Realities and prospects for action in Kwazulu/Natal. *South African Journal of Demography* 3:11–20.

Pritchett, Lant H. 1994. Desired fertility and the impact of population policies. *Population and Development Review* 20 (1): 1–55.

Pullum, Thomas W. 2006. *An assessment of age and date reporting in the DHS surveys, 1985–2003*. DHS Methodological Reports 5. Calverton, MD: Macro International. Available from http://dhsprogram.com/pubs/pdf/MR5/MR5.pdf.

Rabe-Hesketh, Sophia, and Anders Skrondal. 2008. *Multilevel and longitudinal modeling using Stata*. College Station, TX: Stata Press.

RamaRao, Saumya, Marlina Lacuesta, Marilou Costello, Blesilda Pangolibay, and Heidi Jones. 2003. The link between quality of care and contraceptive use. *International Family-Planning Perspectives* 29 (2): 76–83.

Respond Project. 2012. *Making family-planning acceptable, accessible, and affordable: The experience of Malawi*. Project Brief No. 6. New York, NY: USAID.

Ross, John, and Karen Hardee. 2013. Access to contraceptive methods and prevalence of use. *Journal of Biosocial Science* 45 (6): 761–78.

Ross, John, and John Stover. 2013. Use of modern contraception increases when more methods become available: Analysis of evidence from 1982–2009. *Global Health: Science and Practice* 1 (2): 203–12.

Sedgh, Gilda, and Rubina Hussain. 2014. Reasons for contraceptive nonuse among women having unmet need for contraception in developing countries. *Studies in Family-Planning* 45 (2): 151–69.

Sharan, Mona, Saifuddin Ahmed, John May, and Agnes Soucat. 2011. Family-planning trends in sub-Saharan Africa: Progress, prospects, and lessons learned. In *Yes Africa can: Success stories from a dynamic continent*, eds. Punam Chuhan-Pole and Manka Angwafo, 445–63. Washington, DC: The International Bank for Reconstruction and Development/The World Bank. Available from http://siteresources.worldbank.org/AFRICAEXT/Resources/258643-1271798012256/family-planning-25.pdf.

SHOPS Project. 2012. *Malawi private health sector assessment. Brief*. Bethesda, MD: SHOPS Project, Abt Associates.

Skiles, Martha Priedeman, Marc Cunningham, Andrew Inglis, Becky Wilkes, Ben Hatch, Ariella Bock, and Janine Barden-O'Fallon. 2015. The effect of access to contraceptive services on injectable use and demand for family-planning in Malawi. *International Perspectives on Sexual and Reproductive Health* 41 (1): 20–30.

Soler-Hampejsek, Erica, Monica J. Grant, Barbara S. Mensch, Paul C. Hewett, and Johanna Rankin. 2013. The effect of school status and academic skills on the reporting of premarital sexual behavior: Evidence from a longitudinal study in rural Malawi. *Journal of Adolescent Health* 53 (2): 228–34.

StataCorp. 2013. *Stata statistical software: Release 13*. College Station, TX: StataCorp LP.

Stephenson, Rob, Andy Beke, and Delphin Tshibangu. 2008. Community and health facility influences on contraceptive method choice in the Eastern Cape, South Africa. *International Famiily Planning Perspectives* 34 (2): 62–70.

United Nations Department of Economic and Social Affairs Population Division. 2015. World population prospects: The 2015 revision, key findings and advance tables. Working Paper No. ESA/P/WP.241. New York, NY: United Nations.

USAID/Africa Bureau, USAID/Population and Reproductive Health, Ethiopia Federal Ministry of Health, Malawi Ministry of Health, and Rwanda Ministry of Health. 2012. *Three successful sub-Saharan Africa family-planning programs: Lessons for meeting the MDGs*. Washington, DC: USAID.

Wang, Wenjuan, Shanxiao Wang, Thomas Pullum, and Paul Ametepi. 2012. *How family-planning supply and the service environment affect contraceptive use: Findings from four East African countries*. DHS Analytical Studies 26. Calverton, MD: ICF International.

Williamson, Lisa M, Alison Parkes, Daniel Wight, Mark Petticrew, and Graham J Hart. 2009. Limits to modern contraceptive use among young women in developing countries: A systematic review of qualitative research. *Reproductive Health* 6 (3). doi:10.1186/1742- 4755-6-3.

Wood, Kate, and Rachel Jewkes. 2006. Blood blockages and scolding nurses: Barriers to adolescent contraceptive use in south africa. *Reproductive Health Matters* 14 (27): 109–18.

Yao, Jing, Alan T. Murray, and Victor Agadjanian. 2013. A geographical perspective on access to sexual and reproductive health care for women in rural Africa. *Social Science & Medicine* 96:60–68.

Understanding How Low–Socioeconomic Status Households Cope with Health Shocks: An Analysis of Multisector Linked Data

By
TAMMY LEONARD,
AMY E. HUGHES,
and
SANDI L. PRUITT

Low–socioeconomic status (SES) households have little income or wealth to buffer against the negative impacts of adverse health events among adult household members. This research project links data from a nonprofit food distribution center, electronic medical records from a safety-net healthcare system, and publicly available residential appraisals for more than 3,000 households to provide insight into how low-SES households cope with health shocks experienced by resident adults. Three broad types of strategies are examined: changes in household structure, residential mobility, and use of social services. Of the households studied, 20.2 percent had at least one adult member who experienced a health shock. These households were more likely to gain additional adult household members and employed household members, were more likely to move residence and to move distances greater than one mile, and were less likely to visit the food distribution center after the shock. This research highlights how novel data linkages can help us to understand how health and social policies impact vulnerable populations.

Keywords: health shock; food insecure; coping; health disparity

Households and individuals of low socioeconomic status (SES) are more likely to suffer from poorer health and have fewer resources to buffer against the negative effects of poor health (Smith and Kington 1997) than those with high SES. As a result, unexpected adverse

Tammy Leonard is an associate professor of economics at the University of Dallas and an adjunct assistant professor at the University of Texas Southwestern Medical Center. Dr. Leonard is also codirector of the Community Assistant Research (CARE) initiative. She specializes in interdisciplinary applications of public, urban, and behavioral economics.

Amy E. Hughes is a postdoctoral research fellow at the University of Texas School of Public Health. Her research employs spatial statistics to explore the spatial, social, and neighborhood effects of health behaviors.

Correspondence: tleonard@udallas.edu

DOI: 10.1177/0002716216680989

health events can be particularly devastating for low-income households because they can disrupt employment, create new household economic needs (i.e., healthcare costs), and increase household workloads (i.e., providing care for the unhealthy household member). Despite the known challenges for low-income families when household members fall into poor health, relatively little is known about household coping strategies. We know that for higher income families, adverse health events often lead to depletion of savings (Semyonov, Lewin-Epstein, and Maskileyson 2013, Smith 1998, van Doorslaer et al. 1997). However, for households with little or no savings and income, coping strategies are likely to be more diverse.

Because coping strategies of low-SES households have not been explicitly examined, there are few recommendations for how policy-makers or social service organizations might best intervene to provide assistance following adverse health events. This gap in the research is, in part, a result of insufficient data to measure how poor health impacts the lives of those for whom wealth and income were scant prior to an adverse health event. We address this gap by linking robust, unique administrative and clinical datasets to better understand coping strategies employed by low-SES households when an adult household member experiences an adverse health event. Specifically, we relied on unexpected adverse health events (*health shocks*), experienced by adult patients in a safety-net healthcare system as a source of exogenous variation to understand how health impacted household composition changes, employment, housing mobility, and utilization of social services.

Background

Health and household socioeconomic status

Income is negatively correlated with health within every age category (Smith and Kington 1997; Smith 2004), and this correlation is robust across a sample of sixteen developed countries (Semyonov et al. 2013). An association between SES

Sandi L. Pruitt is an assistant professor at the University of Texas Southwestern Medical Center in Dallas. Her research examines geographic, socioeconomic, and racial/ethnic disparities in health behaviors, healthcare outcomes, and healthcare utilization.

NOTE: This work was supported by Robert Wood Johnson Foundation grant 73436, the Program for the Development and Evaluation of Model Community Health Initiatives in Dallas at UT Southwestern Medical Center, and the Agency for Healthcare Research and Quality (R24 HS 22418-01). Dr. Hughes was funded through a postdoctoral fellowship at the University of Texas School of Public Health Cancer Education and Career Development Program, National Cancer Institute/NIH Grant R25 CA57712. We thank Oanh K. Nguyen for helpful comments at all stages of this project, Joanne Sanders for sharing her EMR data expertise, and the leadership and staff at Crossroads Community Services and Parkland Health and Hospital System for partnering with us in this endeavor. This content is solely the responsibility of the authors and does not necessarily represent the official views of the National Cancer Institute or the National Institutes of Health.

and health is common and consistent in numerous studies and across diverse populations, places, and health outcomes. People of lower SES or who live in neighborhoods characterized by lower SES are more likely to have poorer health (Marmot et al. 1991; Blackman and Masi 2006; Drewnowski et al. 2007; Peek et al. 2007).

However, the causal mechanism for this correlation has long been debated (Adler and Ostrove 1999). Evidence from extant research suggests causal pathways from poor health to lower SES (Deaton 2008; van Doorslaer et al. 1997) and vice versa, from low SES to poorer health (Adams et al. 2003; Smith 2005). For example, poor health leads to lower SES if chronic or acute health conditions limit workforce participation and, in turn, this process results in job loss. Low SES leads to poorer health if households have limited access to health resources and exhibit suboptimal health behaviors such as disengaging in preventive health care, avoiding regular doctor visits, or eating unhealthy diets.

The relationship between SES and health also varies across the life cycle. In middle-aged and older adults, poor health is related to lower household SES (Smith and Kington 1997; Smith 1998; Heckman and Smith 1999) because new adult adverse health events lead to less employment, income, and wealth (Smith 2004). Additionally, children who grew up in low-SES households experience poorer health outcomes in adulthood (Smith 2004; Currie 2008; Currie and Almond 2011). Household SES particularly impacts the health of household members during childhood and early adulthood when income and wealth trajectories are being established (Condliffe and Link 2008). Thus, when an adult household member experiences poor health, household SES may decline. This decline in household SES may in turn impact the health of children in the household and their long-term health trajectories. Therefore, the effects of poor health and low-SES are compounded within the household.

Adverse health events and health shocks

Understanding the frequency and impact of adverse health events experienced by adult household members is critical to designing strategies to improve the well-being of all individuals living in low-SES households. Adverse health events exist along a continuum based on the degree to which they can be anticipated and prevented. For example, adverse health events resulting from natural disasters (e.g., tornado, earthquake, tsunami) are largely unpreventable and may be generally considered exogenous to other individual preventive health behaviors. However, hypoglycemia in a diabetic patient caused by failure to take prescribed medication or follow a prescribed diet may be considered highly preventable and is endogenously related to the patient's behavior. In the medical literature, adverse health events such as hospitalizations are characterized as either preventable and nonpreventable based on whether access and utilization of primary care could have theoretically prevented the hospitalization (Parchman and Culler 1999).

Researchers have utilized adverse health events as an exogenous source of variation to identify the causal impact of adverse health events. These events, often referred to as *health shocks*, produce a quasi-experimental framing: the treatment group consists of patients who experience a health shock and the control group consists of patients who do not experience a health shock. However, this causal identification strategy relies on correctly characterizing health shocks such that they are truly accounted for as exogenous events (Smith 2004).

While researchers largely agree on the validity of the quasi-experimental design, there is significant heterogeneity in how health shocks are defined. For example, studies utilizing data from the Health and Retirement Study or the Panel Study of Income Dynamics have defined health shocks as onset of acute or chronic conditions such as heart problems, stoke, cancer, lung disease or diabetes, and/or changes in self-rated health or hospitalizations (Wu 2003; Berkowitz and Qiu 2006; Condliffe and Link 2008; Lee and Kim 2008; Conley and Thompson 2011; Bradley et al. 2012; Kim, Yoon, and Zurlo 2012). Other studies have used databases such as the National Highway Traffic Safety Administration to identify car crash victims (Doyle 2005). Others have analyzed household survey data to study change in self-reported health status or incidence of chronic physical or mental health problems (García-Gómez 2011) as a measure of health shocks. Clearly, these characterizations of health shocks differ in terms of the extent to which the adverse health events may be considered strictly exogenous, and this is a known limitation of the literature.

Coping strategies

Despite this limitation, robust evidence suggests that health shocks tend to be most severe among low-SES households and those without private health insurance (Berkowitz and Qiu 2006; Lee and Kim 2008; Conley and Thompson 2011; Kim et al. 2012, Curtis et al. 2013). However, we know little about how low-SES households who have little income or wealth cope with health shocks. Very low-SES households are underrepresented in previous work because the literature has largely focused on the impact of health shocks on income and wealth. Moreover, many prior survey studies face limitations in recruitment of these populations. Some studies have examined coping strategies among the food insecure, a particular subset of the low-SES population. The food insecure are individuals or households without reliable access to a sufficient quantity of affordable, nutritious food.

Studies suggest that adverse health events can impact low-income households far beyond depletion of household financial resources. An adverse health event to an adult family member may cause job loss, increased expenses for healthcare, and high recurring expenses to obtain medication (Smith 2004). Additionally, households experiencing these challenges are also more likely to report food insecurity (Edin et al. 2013; Noonan, Corman, and Reichman 2014). In qualitative work, researchers have found that prior to cutting back on food, household members engage in multiple coping strategies. These included relying on social networks to meet family needs, turning to nonprofit food distributors, and engaging in strategic shopping patterns to stretch limited financial resources (Edin et al. 2013).

Finally, researchers have also documented coping strategies used by households that experience income volatility. These strategies include accessing savings, increasing utilization of credit, and relying on public benefit programs (Andersen et al. 2015). Notably, low-SES households engage in these coping strategies in creative ways. For example, households with little savings of their own often access the savings of friends or family members, seek help from informal savings circles (especially among immigrant populations), or make early withdrawals from retirement savings accounts. Additionally, access to credit from traditional sources can be limited for low-SES households, thus they often turn to nontraditional sources such as payday lenders (Morduch, Ogden, and Schneider 2014).

Across multiple domains, the extant literature suggests that key coping strategies are related to either seeking help among family and friends or seeking help from social service organizations. Empirically, we might observe coping strategies related to seeking help from family and friends by observing changes in household composition. Coping strategies related to seeking help from social service organizations may be observed on both the number of visits to service providers and the quantity of help received during a visit. Finally, we might observe increases in food insecurity or changes in residential address as either a failure of the primary coping strategies or as a coping strategy itself. For example, a household member may report food insecurity because the family is cutting meal size to devote resources to improving health of the sick adult wage earner; or a family may cut meal size because they have exhausted all other possible alternatives (Edin et al. 2013). Similarly, household members may double up—that is, move in with extended family—to provide extra help with household caregiving; or they may move in with extended family because of eviction (Ahrentzen 2003).

Data and Methods

Data sources

To estimate the impact of health shocks on policy-relevant outcomes, we linked three datasets: (a) Crossroads Community Services administrative data, Parkland Health and Hospital System; (b) electronic medical record (EMR) data; and (c) Dallas Central Appraisal District (DCAD) housing appraisal data.

Crossroads Community Services (>5,500 households annually). Crossroads Community Services (hereafter "Crossroads") is the largest nonprofit food distributor in Dallas County, Texas. It distributed more than 2.6 million pounds of food to 15,787 individuals in 2014. Crossroads maintains a robust, longitudinal database comprising data from their low-SES (<185% federal poverty line) clients. The database includes client- and household-level measures of demographic and socioeconomic information, residential location, food selection, and service utilization.

Parkland EMR (~65,000 records annually). Parkland Health and Hospital System (hereafter "Parkland") is one of the largest integrated safety-net health care systems in the United States, reporting 1.5 million patient encounters annually. Parkland consists of a 900-bed hospital, specialty clinics, and 12 community-oriented primary care clinics that collectively provide comprehensive inpatient, outpatient, specialty, and primary care for under- and uninsured, low-income Dallas County residents. Parkland uses the state-of-the-art comprehensive system-wide Epic EMR system (Epic Systems Inc., Verona, WI) to track its patients. Parkland EMR data include patient-, provider-, clinic-, and system-level data incorporating both medical and nonmedical data (e.g., demographic, health conditions, medical procedures and tests, and healthcare utilization data). Because Parkland is an integrated health system and the only safety-net provider in Dallas County, it has nearly complete coverage for low-income, under- or uninsured adult county residents.

Dallas Central Appraisal District Data (>650,000 parcels). Publicly available appraisal data (appraisal value, home characteristics, and parcel location) for Dallas County, Texas, were obtained from the local tax authority, the DCAD. Appraisal data are recorded annually for each housing parcel ($n > 630{,}000$).

Sample

The analytic sample includes all households in the Crossroads administrative database with: (1) more than one Crossroads visit occurring in the two-year study window of December 1, 2013–November 30, 2015; (2) Crossroads visits spanning a minimum of 180 days; and (3) at least one household member who was an established Parkland patient. Households with one established Parkland patient were defined as having at least one inpatient or outpatient encounter with Parkland between 2004 and 2015.

Variables

Dependent variables, derived using Crossroads and DCAD data, measure three distinct types of household coping strategies as identified in the extant literature: utilization of Crossroads food assistance, household composition changes, and changes in residential address. Our measure of Crossroads utilization is the Crossroads visit rate, defined as the number of visits to receive charitable food assistance within the study window divided by the number of months in the study window. Measures of household composition changes included a change in the number of adults in the household and whether a household gained an employed (full or part time) adult. Households were characterized as gaining an employed adult if they both gained an additional adult household member and there was an additional adult household member employed. Measures of residential address change included whether the household changed residence and whether a household moved a long distance (more than one mile). Moves were determined by comparing geocoded addresses at the first and last Crossroads visit in the study window.

Our primary independent variable—the presence of a health shock between Crossroads visits—was obtained from the Parkland EMR. Health shocks occurring between the first and last Crossroads visit in the study window were identified using EMR data. We modeled the effect of health shocks on dependent variables by considering occurrence of a health shock to any adult household member. Separately, we modeled health shocks occurring to the head of household and to other adult household members (nonhead of household) among a subsample of households with >1 adult household member (hereafter, the "multiple-adult subsample"). The head of the household was defined as the client who is the primary household member responsible for selecting and transporting food for the household during visits to Crossroads.

To generate unbiased estimates of health shock effects on subsequent behavior, we identified exogenous shocks that were likely to be uncorrelated with changes in patient behavior (e.g., sudden uptake of preventive services). To satisfy these restrictions, we limited our characterization of health shocks to emergency department visits for clinical encounters (e.g., not for the purpose of medication refills) for conditions that are not potentially preventable. We excluded potentially preventable visits using diagnosis codes (i.e., ICD-9), following a standardized approach common in the health services literature (Parchman and Culler 1999). Following this approach, preventable conditions include, for example, asthma or hypertension (which can be managed by patient adherence to maintenance therapies).

Other covariates were obtained from Crossroads and Parkland EMR data. Covariates drawn from Crossroads data included household size, presence of children (<18 years of age) in the household (yes/no), number of household members employed full or part time (continuous); and for the head of household: age (continuous), sex (male/female), race/ethnicity (Hispanic, black, white/other), marital status (no/yes), and highest level of education obtained (some high school, high school graduate, college graduate). We controlled for the length of time between the two Crossroads visits during which a health shock could have occurred. EMR covariates included the total number of outpatient visits (between 2012 and 2015) and type of healthcare insurance (uninsured/charity care, Medicaid, Medicare, other/unknown). Presence of children, employment, and marital status were measured based on the values provided at the first Crossroads visit within the study window.

Analytic methods

Linking administrative datasets. Address histories for each of the eligible households were geocoded following a hierarchical geocoding process comprising three levels: (1) parcel, or cadastral, information (Murray et al. 2011); (2) ESRI's StreetMap Premium product (ESRI and Tele Atlas 2012); and (3) Google application program interface (API). Each address was passed to the cadastral address locator first, and the unmatched addresses from the cadastral-level geocoding results were passed to the StreetMap address locator. Any unmatched addresses from the street-level geocoding results were then passed to the Google

API for the final geocoding. A total of 8,165 addresses were geocoded; 63.2 percent ($N = 5,159$) matched at the cadastral level, 29.4 percent ($N = 2,399$) matched at the street level, and 7.4 percent ($N = 607$) matched with the Google API.

The second and third geocoding levels produce coordinates that correspond to the middle of streets; these coordinates were adjusted to the nearest parcel. Some coordinates ($N = 746$) corresponded to parcels associated with more than one DCAD account, and subsequently, more than one parcel value. For these parcels, Crossroads addresses were matched to a single DCAD account if text from the second line of the address provided adequate unit-level information to match a DCAD account associated with a unit. If the second line of the address did not provide adequate unit-level information, then the account with the highest property value was chosen. For accounts with equal value, the account with the most appropriate building (e.g., residential) description was chosen. For the analytic sample, 100 percent of the provided addresses were matched to a DCAD parcel through one of the three hierarchical geocoding levels.

We matched Crossroads data to EMR data using the following criteria: (1) 100 percent match on date of birth and (2) fuzzy match on patient name. The 'STRINGDIST' command in R was used to match first and last name, or to match a similar full name (van der Loo 2014). If two or more Parkland patients matched on date of birth and fuzzy name, then ties were broken by a 100 percent match on zip code (i.e., if any of the EMR zip codes matched any of the Crossroads zip codes).

The analytic sample included all households represented in the Crossroads administrative database from December 1, 2013, through November 30, 2015. The initial Crossroads data included 71,349 visits to a food distributor by 10,840 unique individuals residing in 7,588 households. Only records with more than one Crossroads visit, and with a Crossroads history spanning more than 179 days, were included in the analytic sample to facilitate measurement of change in behavior associated with coping strategies. The final Crossroads sample comprised 4,471 households. Of these, 3,696 (83 percent) were engaged in the Parkland Health System, defined as having at least one adult household member with a Parkland encounter between 2004 and 2015. One hundred percent of the matched household addresses were geocoded to a DCAD parcel using the methods described previously. Of the 3,696 households that could be matched across all three administrative datasets, 3,235 had complete data for all variables used in the analysis that follows. Thus the sample of 3,235 households comprises our analytic sample.

Multivariate analysis. We estimated three multivariate regression models for each dependent variable. The models differed in terms of how the health shock variable was defined. We examined household-adult health shocks in the full analytic sample by defining the health shock as one or more health shocks experienced by any adult household member during the study window (hereafter a "household-adult health shock"). In the multiple-adult subsample, the health shock variable was defined differently in two separate models. In the first multiple-adult subsample model, the health shock variable was defined as one or more health shocks experienced by the adult head of household during the study window (hereafter a "head health shock"). In the second multiple-adult subsample model, the health shock variable was defined as one or more health shocks experienced by a nonhead adult household member during the study window (hereafter "nonhead health shock"). We examined head and nonhead health shocks only in the multiple-adult subsample to facilitate a comparison of how households cope when different types of adult members experience poor health.

Crossroads visiting frequency was modeled as a continuous dependent variable and parameter estimates were obtained using ordinary least squares regression. Ordered logistic regression was used to estimate parameters for models with change in number of adult household members as the dependent variable. Finally, logistic regression models were estimated for all binary dependent variables.

Results

Descriptive analysis

Summary measures for variables included in the linked analytic sample are provided in Table 1. Household size varied from one to fourteen members and included zero to three employed adults. More than one-third (37 percent) included children. For each household, the individual who most often visited Crossroads to obtain food was identified as the head of household. Average head of household age was 47 years, and 32 percent were married. Half (51 percent) of the head of households have a high school degree or more education. Thus, 49 percent of the households in our sample have a head who has less than a high school education. The sample was primarily non-Hispanic black (53 percent) or Hispanic (33 percent).

Household-adult health shocks occurred in 20 percent of the full sample. Most households lacked insurance (52 percent). The average Crossroads utilization rate was 0.85; in other words, households visited Crossroads for food slightly less than once per month. In the full analytic sample, 38 percent of households had a Crossroads utilization rate of one or greater, indicating that they visited Crossroads at least once per month during the study window. On average, household health system participation amounted to twenty-six outpatient encounters per household between 2012 and 2015. Among households that experienced changes in the number of adult members, more households gained (12.6 percent) than lost (3.4 percent) adult members. Additionally, 3.7 percent of the sample gained an employed adult member, and 8.2 percent moved during the study window.

Table 1 describes the multiple-adult subsample ($n = 1,151$). For this subsample, head health shocks occurred in 16.9 percent of households and nonhead health shocks occurred in 12.2 percent of households. In comparison with the full

TABLE 1
Summary Statistics, Full Analysis Sample and Multiple-Adult Sample

	Full Sample	Multiple-Adult Sample
	$N = 3{,}235$	$n = 1{,}151$
Continuous variables		
	Mean (SD)	Mean (SD)
Crossroads visit rate	0.85 (0.26)	0.78 (0.28)
Study window (days)	501.07 (178.1)	517.49 (175.85)
Household size	2.53 (1.99)	4.19 (1.93)
Employed adults	0.26 (0.49)	0.55 (0.62)
Head's age	47.43 (17.74)	43.03 (13.99)
Health visits (any adult household member)	25.92 (40.02)	22.93 (35.16)
Discrete variables		
	Percent	Percent
Health shock to head	20.2	16.9
Health shock to nonhead	—	12.2
Children in household	37.3	63.9
Male head of household	24.2	15.8
Married head of household	31.7	65.9
Household gained an employed adult	3.7	3.9
Moved	8.2	9.6
Moved > 1 mile	5.7	6.9
Head's race/ethnicity		
Non-Hispanic white	12.9	9.2
Hispanic	33.3	61.1
Non-Hispanic black	52.9	28.9
Head's educational attainment		
Completed high school (but no college)	37.9	30.7
Some college (or more)	13.6	9.3
Head's insurance status		
No insurance	51.8	67.4
Medicaid	17.4	10.5
Medicare	26.8	16.0
Other insurance	3.9	6.1
Change in number of adults in household		
Lost > 1 adult	0.7	1.8
Lost 1 adult	2.9	8.1
No change	83.3	74.3
Gained 1 adult	10.4	12.9
Gained > 1 adult	2.2	3.0

analytic sample, households in the subsample are larger and the proportion of households with children is higher. The mean age of household head was comparatively younger, and the subsample included more Hispanic households (61 percent) than non-Hispanic black (29 percent) households. Heads of households in the subsample were less educated than in the full sample. Within the multiple adult household sample, there were slightly more changes in the number of adults in the household during the study period and residential mobility was slightly higher (9.6 percent of households moved). Crossroads utilization rate and health system participation (i.e., number of medical encounters) were lower for the subsample.

Multivariate regression results

Table 2 presents estimates for multivariate least squares regression models examining the impact of health shocks on Crossroads visiting rate. A household-adult health shock in the full sample model (first column, Table 2) is associated with a lower Crossroads utilization rate, but has no significant association with the Crossroads utilization rate in the multiple-adult household subsample (second and third columns, Table 2). Across all models, households with an older head and a higher number of employed adults visited Crossroads more frequently, while Hispanic households visited less frequently than non-Hispanic households. Additionally, in the household-adult shock model, households with a married head of household visited less frequently.

The relationship between health shocks and household compositional changes were measured by examining (1) changes in the number of adults in the household via ordered logistic regression and (2) additions of an employed adult household member using logistic regression (Table 3). A household-adult health shock in the full sample (first 2 columns, Table 3) and a nonhead health shock in the multiple-adult subsample (columns 5 and 6, Table 3) are associated with both an increase in the number of adults in the household and increased odds of gaining an employed adult household member. The effect sizes are noteworthy, particularly when examining a health shock to a nonhead adult household member: households are nearly twice as likely to gain any adults and to gain employed adults when there is a health shock to a nonhead adult household member. Interestingly, there are no statistically significant relationships between household composition and head health shocks.

Other strong correlates of changes in household composition include the presence of children in the household and race/ethnicity (Table 3). In the household-adult shock models, households are 4.5 times more likely to add adult household members and twice as likely to add an employed adult if there are children present. Considering only the multiple-adult subsample, when a shock occurred to a head or nonhead, households were twice as likely to add adult household members when there were children, but there is no significant relationship between presence of children and adding employed adults. In contrast, race/ethnicity was more strongly associated with adding employed adults. Compared to non-Hispanic households, Hispanic households were nearly eight (full sample) or five (multiple-adult subsample) times more likely to add employed adults.

TABLE 2
Ordinary Least Squares Estimation Results for Crossroads Utilization Rate

	Full Sample	Multiple-Adult Subsample[a]	
	Household-adult shock	Head shock	Nonhead shock
Health shock	−0.043***	−0.019	0.021
	(0.011)	(0.021)	(0.024)
Children in household	−0.031+	−0.03	−0.03
	(0.017)	(0.025)	(0.025)
Study window	0.000	0.012	0.011
	(0.004)	(0.008)	(0.008)
Household size	−0.008+	0.003	0.003
	(0.004)	(0.006)	(0.006)
Employed adults	0.045***	0.028+	0.029*
	(0.01)	(0.014)	(0.014)
Head's age	0.002***	0.003***	0.003***
	(0.000)	(0.001)	(0.001)
Hispanic	−0.142***	−0.15***	−0.15***
	(0.013)	(0.023)	(0.023)
Male head of household	−0.019	0.02	0.019
	(0.011)	(0.022)	(0.022)
Completed high school	0.018	0.014	0.014
	(0.01)	(0.019)	(0.019)
Some college	−0.014	−0.017	−0.016
	(0.014)	(0.028)	(0.028)
Married head	−0.036**	−0.005	−0.005
	(0.012)	(0.019)	(0.019)
Health visits	0.000	0.000	0.000
	(0.000)	(0.000)	(0.000)
Medicaid	−0.001	0.061*	0.06*
	(0.013)	(0.026)	(0.026)
Medicare	0.000	0.006	0.006
	(0.012)	(0.026)	(0.026)
Other insurance	0.002	−0.011	−0.01
	(0.022)	(0.032)	(0.032)
Constant	0.849***	0.754***	0.749***
	(0.025)	(0.049)	(0.049)
Observations	3,235	1,151	1,151
R^2	.17	.14	.14

***$p < .001$. **$p < .01$. *$p < .05$. +$p < .10$.
NOTE: Standard errors in parentheses.
a. The multiple-adult subsample is defined as all households having > 1 adult member.

TABLE 3
Estimated Odds Ratios for Models Examining Change in Household Composition

	Full Sample		Multiple-Adult Subsample[a]			
	Household-adult shock		Head shock		Nonhead shock	
	Change in number of adults in household[b]	Household gained an employed adult[c]	Change in number of adults in household[b]	Household gained an employed adult[c]	Change in number of adults in household[b]	Household gained an employed adult[c]
Health shock	1.630°°°	1.800°	1.308	1.153	1.974°°°	2.048+
	(0.196)	(0.431)	(0.239)	(0.481)	(0.395)	(0.846)
Children in household	4.559°°°	2.066+	2.078°°	2.095	2.114°°°	2.13
	(0.838)	(0.85)	(0.472)	(1.396)	(0.481)	(1.412)
Study window	1.113°	1.031	1.074	0.968	1.072	0.944
	(0.055)	(0.112)	(0.074)	(0.158)	(0.074)	(0.156)
Household size	0.768°°°	0.910	0.897°	1.122	0.890°	1.120
	(0.036)	(0.067)	(0.048)	(0.113)	(0.048)	(0.113)
Employed adults	0.526°°°	0.131°°°	0.795+	0.223°°°	0.784+	0.216°°°
	(0.062)	(0.034)	(0.104)	(0.074)	(0.102)	(0.072)
Head's age	0.993	0.954°°°	0.995	0.981	0.994	0.98
	(0.005)	(0.011)	(0.007)	(0.017)	(0.007)	(0.017)
Hispanic	1.397°	7.678°°°	1.114	4.843°	1.088	4.805°
	(0.208)	(2.899)	(0.225)	(3.071)	(0.221)	(3.04)
Male head of household	0.837	0.306°	0.994	0.192	0.975	0.176+
	(0.106)	(0.165)	(0.19)	(0.2)	(0.187)	(0.184)
Completed high school	0.857	0.780	0.886	0.884	0.879	0.900
	(0.097)	(0.204)	(0.144)	(0.378)	(0.143)	(0.387)
Some college	0.796	0.806	0.738	1.845	0.72	1.834
	(0.128)	(0.358)	(0.185)	(1.095)	(0.181)	(1.095)
Married head	1.615°°°	2.729°°°	1.122	1.201	1.098	1.172
	(0.213)	(0.714)	(0.183)	(0.494)	(0.179)	(0.479)
Health visits	0.999	0.991	1.000	0.996	1.000	0.995
	(0.001)	(0.005)	(0.002)	(0.007)	(0.002)	(0.007)
Medicaid	0.820	0.740	0.824	0.379	0.818	0.376
	(0.119)	(0.263)	(0.189)	(0.289)	(0.188)	(0.287)
Medicare	0.876	0.794	0.821	0.924	0.829	0.925
	(0.129)	(0.424)	(0.187)	(0.661)	(0.189)	(0.666)
Other insurance	1.099	0.742	1.016	0.574	1.069	0.612
	(0.261)	(0.325)	(0.29)	(0.432)	(0.306)	(0.461)
Observations	3,235	3,235	1,151	1,151	1,151	1,151

°°°$p < .001$. °°$p < .01$. °$p < .05$. +$p < .10$.
NOTE: Standard errors in parentheses.
a. The multiple-adult subsample is defined as all households having >1 adult member.
b. Ordered logistic regression results reported as odds ratios.
c. Logistic regression results reported as odds ratios.

Health shocks are also associated with change in residential address (Table 4). In the full sample, households experiencing a health shock were two times more likely to move and to move more than one mile compared with households that did not experience a health shock. In the multiple-adult subsample, estimated effect sizes are similar when a health shock was experienced by the head of household, but parameter estimates are not statistically significant in models examining nonhead health shocks. Most other covariates are not related to residential address changes, with the exception of household head age (among all models, decreased likelihood of moving and moving long distances), race/ethnicity (for the full sample only; Hispanic households are more likely to move, but not move long distances), household size (for the full sample only; larger households are more likely to move and move longer distances), and employed adults (for the full sample only; households with more employed adults are less likely to move and move longer distances). Finally, among all models, those with longer study windows were more likely to move.

Discussion

We found that household-adult health shocks are related to a number of coping behaviors, including decreased Crossroads utilization, changes in household composition (i.e., an increase in adults and an increase in employed adults), and an increased likelihood of residential moves as well as moving longer distances. Additionally, our results are suggestive of interesting correlates between coping behaviors and household demographics that likely can influence the direction of future research agendas. Hispanic households and households with children are much more likely to add adults and add employed adults when a health shock occurs.

When comparing results for head versus nonhead health shocks in the multiple-adult subsample, we found evidence for coping strategies that vary depending on which household adult endured the health shocks. When nonhead adults experience health shocks, the household is more likely to "double up"—take on additional adult members to distribute expense and care burdens. When the head of the household's health is affected, households are more likely to move and move longer distances.

Value of multisector data linkages

Our analysis demonstrates how linked data from the social service, health, and housing sectors can improve our understanding of how households manage food, employment, and housing resources to cope with an adverse health event. Previous work in this area has relied on changes in income or wealth (Wu 2003; Berkowitz and Qiu 2006; Condliffe and Link 2008; Lee and Kim 2008; Kim et al. 2012). Wealth and income measures may be less relevant for understanding the impacts of health shocks for our low-SES population, who are largely

TABLE 4
Estimated Odds Ratios for Models Examining Change in Residential Address

| | Full Sample | | Multiple-Adult Subsample[a] | | | |
| | Household-adult shock | | Head shock | | Nonhead shock | |
	Moved	Moved > 1 mile	Moved	Moved > 1 mile	Moved	Moved > 1 mile
Health shock	1.992°°°	2.219°°°	1.838°	1.926°	0.981	1.199
	(0.292)	(0.373)	(0.459)	(0.55)	(0.313)	(0.416)
Children in household	0.87	0.71	0.57	0.459+	0.571	0.46+
	(0.211)	(0.202)	(0.205)	(0.187)	(0.205)	(0.187)
Study window	1.307°°°	1.159+	1.199+	1.076	1.233+	1.103
	(0.092)	(0.094)	(0.132)	(0.133)	(0.135)	(0.136)
Household size	1.107°	1.184°°	1.13+	1.182°	1.128+	1.179°
	(0.056)	(0.067)	(0.076)	(0.091)	(0.076)	(0.09)
Employed adults	0.7°	0.668°	0.892	0.93	0.849	0.886
	(0.103)	(0.117)	(0.172)	(0.205)	(0.162)	(0.193)
Head's age	0.955°°°	0.955°°°	0.952°°°	0.961°°	0.951°°°	0.959°°°
	(0.006)	(0.007)	(0.011)	(0.012)	(0.011)	(0.012)
Hispanic	1.81°°	1.272	1.545	1.471	1.542	1.452
	(0.362)	(0.304)	(0.52)	(0.567)	(0.514)	(0.554)
Male head of household	1.102	1.239	0.733	0.91	0.738	0.908
	(0.201)	(0.258)	(0.269)	(0.358)	(0.269)	(0.354)
Completed high school	0.871	0.988	1.043	1.171	1.045	1.174
	(0.138)	(0.182)	(0.26)	(0.335)	(0.259)	(0.335)
Some college	1.245	1.295	1.312	1.866	1.264	1.788
	(0.263)	(0.322)	(0.506)	(0.77)	(0.483)	(0.731)
Married	0.747+	1.004	0.63+	0.758	0.61+	0.724
	(0.131)	(0.205)	(0.162)	(0.225)	(0.156)	(0.214)
Health visits	0.996	0.998	0.996	0.997	0.997	0.998
	(0.002)	(0.003)	(0.004)	(0.005)	(0.004)	(0.005)
Medicaid	1.242	1.136	0.747	0.57	0.765	0.584
	(0.232)	(0.248)	(0.277)	(0.261)	(0.283)	(0.267)
Medicare	1.466+	1.322	1.25	0.856	1.248	0.866
	(0.319)	(0.335)	(0.507)	(0.413)	(0.504)	(0.418)
Other insurance	1.095	0.87	1.025	0.826	1.059	0.869
	(0.336)	(0.337)	(0.417)	(0.406)	(0.43)	(0.427)
Constant	0.398°	0.233°°°	0.656	0.267+	0.808	0.339
	(0.148)	(0.101)	(0.43)	(0.201)	(0.526)	(0.253)
Observations	3,235	3,235	1,151	1,151	1,151	1,151

°°°$p < .001$. °°$p < .01$. °$p < .05$. +$p < .10$.
NOTE: Standard errors in parentheses.
a. The multiple-adult subsample is defined as all households having >1 adult member.

unemployed and have very low income. People of higher wealth and income (even marginally), who are not in this sample, have the luxury of spending wealth or income *before* they need to do things like seek social services, double up, or move.

Linked data from multiple safety-net systems improve equitable representation in research and reduce the research burden for vulnerable populations. Low-SES populations frequently provide an array of data to social service providers, yet are often underrepresented in research. For example, many Crossroads clients have filled out extensive applications to qualify for a host of social services (e.g., Women, Infants, and Children, Supplemental Nutrition Assistance Program, Social Security, Section 8 housing); middle- and upper-income families are not burdened with providing such information. Many agencies require routine recertification wherein data are collected longitudinally. In addition, nonprofit agencies are increasingly collecting or sharing additional data from clients in order to fulfill funder obligations or to meet criteria for claiming evidence-based programming. Increased utilization and linkage of these data can improve representation of low-SES populations in research. Moreover, utilization of EMR data to ascertain health status bypasses the use of respondent surveys for primary data collection. Respondent surveys entail significant response burden, can be adversely impacted by low response rates and bias, and are often cost-prohibitive.

Utilizing administrative nonprofit data provides significant advantages for researchers wishing to understand vulnerable populations. Linking EMR data to additional sources of social, behavioral, and economic data has tremendous potential for transdisciplinary health research. Parallel efforts in the health services literature have long acknowledged the power of EMR data linkages to improve health outcomes research (Bradley et al. 2010). However, these efforts have been characterized as technologically challenging and hard to manage due to the required interfacing between multiple independent organizations, often with competing needs, regulatory rules, and goals (Bradley et al. 2012). Our work suggests that these challenges may be overcome through fostering collaborative multisector relationships. These relationships may be centered on shared goals and mutual benefits resulting from greater data integration. Widespread adoption of EMRs in the United States will make future multisector data linkages with healthcare partners potentially feasible across many communities.

Nonprofit organizations also benefit from nonprofit data linkages. They can benefit from long-term collaboration with academic researchers, for example. Our study is an example of the benefits that such a long-term investment can yield. The study authors have formed the Community Assistance Research (CARE) initiative to work alongside nonprofit agencies to increase their capacity for collecting and learning from their administrative data. In turn, nonprofits can leverage newly created data to evaluate service provision; report quality metrics and benchmarks to funders; and guide future innovations in program development, implementation, and evaluation.

Multisector linked data analysis presents advantages over alternative research designs. Significant investments of time and relationship-building are needed to establish trust to recruit participants for traditional cross-sectional and cohort research designs. Moreover, in these designs, response rates—particularly for low-SES populations—are often low and attrition is high. Administrative data, in contrast, can be embedded in service provision, and response rates can be exceptionally high. The robust data collection at Crossroads highlighted herein leverages the long-established trust that the organization has with their clients to ensure data collection efficiency and quality. Notably, many administrative data sources present an additional advantage in that they are continually collecting data. In our linked data, DCAD housing appraisal data are released yearly and Crossroads administrative and Parkland EMR data continue to accrue. Thus data are available to researchers on a rolling basis.

Fostering transdisciplinary research with multisector data linkages

While each stakeholder may have unique applications for the linked data, the data linkage endeavor in itself is a shared goal and can provide incentive for all parties to work together. In turn, engaging multiple stakeholders in the data collection, analysis, linkage, and sharing process ensures development of meaningful data that can be deployed by all partners. In our case, CARE has served as a conduit between institutions and between researchers from diverse backgrounds. With the assistance of academic researchers, the nonprofit agencies affiliated with CARE have increased their capacity for collecting and learning from their administrative data. Additionally, with the assistance of nonprofit leaders, the researchers affiliated with CARE have developed more actionable research agendas.

Our study benefits from strengths drawn from public health, economics, and an understanding of the situations faced by low-income families. Economics literature provides a robust framework for characterizing exogenous health shocks; the public health literature allows for an understanding of health services in order to apply the exogeneity criteria to identify health shocks within EMR data; and our food bank partner provided the rationale for collecting many of the coping outcome measures that were examined. Research results generated from the work presented here demonstrate how the nonprofit and academic research sectors can work together to generate policy-relevant, actionable results facilitated by novel data linkages.

Implications for policy and nonprofit service providers

Our results raise numerous questions that are of high importance for public policy design and nonprofit service provision. The absence of studies examining the downstream impacts of adverse health events for low-income populations is in itself a finding of significant concern. While multisector administrative data linkages and collaborations between researchers and nonprofit service providers are long-term goals that hold promise for improving the social safety net, there are intermediate steps that might be taken by the policy and service provider sectors.

First, public policy can better help to support the development of robust systems to facilitate data linkages. Currently, data linkages are technically challenging, and privacy concerns often present additional challenges. However, data linkages have been successfully implemented in some cases. This suggests that improved dissemination and sharing of best practices and technological innovation in this area would provide numerous benefits. Additionally, our work highlights the valuable role of nonprofit service providers as critical data contributors. However, nonprofits struggle with the challenges of investing in the robust data collection systems and the people who can operate them effectively. Funding mechanisms that support long-term technology development, implementation, and maintenance will improve the capacity of nonprofits to leverage and contribute to the benefits of technology investments being made in the private and government sectors.

Additionally, the results presented here point to important areas that nonprofit service providers can begin to engage in prior to any robust data linkages. For example, we report numerous coping mechanisms that occur following a health shock. Results suggest that nonprofit service providers might improve services by systematically asking clients of any health system engagement at each visit. Additionally, results suggest that low-income families change their behavior in multiple dimensions following an unexpected health shock. For instance, they move, change patterns of access to social services, and change the composition of the household. This suggests that recent trends in building partnerships between nonprofits specializing in diverse service provision have the potential to serve a viable roll in intervening in times of household crisis. Information sharing within nonprofit partnership networks may further improve the impact of these collaborations. Thus, data linkages are important both across diverse sectors (i.e., health systems, nonprofits, housing system, and so on) and within sectors (i.e., among nonprofit service providers).

Limitations

Our study has a number of limitations, many of which are reflective of potential shortcomings from our linked data sources. Crossroads data collection is opportunistic, which generated a nonrandom sample and nonrandom timing of data collection. This limits the external validity of our results. We also employed a rather nontraditional characterization of head of household based upon the definition used by Crossroads. Low-SES households have very diverse household structures, rendering any universal definition of head of household problematic. Our results should be interpreted in light of how head of household has been defined within this study—the adult household member who typically acquires food for the household. Finally, we were not able to exhaustively examine all coping strategies. In particular, we did not observe changes in social service utilization in terms of number of services obtained at each visit because Crossroads limits the amount of food and services given. Additionally, we did not have data on food security of the households in our study or measures of direct cash transfers that the household may have received from family or friends.

The use of EMR data to identify health shocks has some advantages and disadvantages. Use of EMR data to ascertain health shocks is an advantage over the majority of the health shocks literature to date, which has largely defined health shocks using self-reported survey data. In contrast to survey data, which suffer from recall and mono-method biases, EMR data provide an objective source of health data wherein health events are date- and time-stamped. However, the exogeneity of health shocks cannot be assumed based on review of ICD-9 codes of emergency room visits. Some included codes may represent conditions that are related to underlying client behaviors, such as a tendency to engage in more risky behavior. This possibility may create confounding and challenge the causal interpretation of our results. Additionally, EMR data are limited to only a single healthcare system. However, it is the primary integrated safety-net health system in the region, so exclusive utilization of this system by our sample is expected to be very high. Nevertheless, some health shocks may not be observed if patients choose to visit a healthcare provider outside of the Parkland system or choose not to seek healthcare.

Data sustainability and future directions

The data linkages leveraged by this study are part of a larger effort to establish a longitudinal data resource for diverse users: the Hunger Center Longitudinal Database (HCL Database). De-identified Crossroads data are archived annually to build the HCL Database. This endeavor is self-sustaining due to a collaboration among CARE, Crossroads, and the regional food bank. CARE guides the scientific rigor and validity of Crossroads data collection through new question development, data collection process monitoring, and evaluation of existing data. Crossroads facilitates fidelity in data collection procedures, ensures client representation, and advocates for innovative use of the data to benefit clients. The regional food bank maintains data sharing policies and makes data available to the research and nonprofit community. To date, two years of data have been archived in the HCL Database.

The HCL Database was designed to create, house, and share administrative data that are scientifically valid and amenable to linkage with other data sources. Because none of the data collection used in this study is tied to a funding stream (e.g., research grant), we anticipate data collection to continue indefinitely. We have facilitated the appropriate memorandums of understanding to support data storage and administration of sharing agreements among the local nonprofit community.

Our study highlights several important areas for further research using multi-sector data linkages, such as the linkage highlighted herein. First, the impact of health shocks on very low-SES households needs to be further examined in light of both (1) a broader array of coping strategies (e.g., food security, more refined financial information) and (2) a more granular examination of the type, severity, and time horizon of health shocks (e.g., trauma, disability). Second, our study highlights the role of residential mobility in households' coping behaviors. This warrants further study and will inform our understanding of how, when, and why neighborhood social and built environments impact low-income households.

Conclusions

Health shocks impact low-income households in ways that result in changes in household structure, mobility, and utilization of social services. Interestingly, for the low-income sample examined here, employment was less likely to be affected, perhaps because employment was already fragile for these households even prior to a potential health shock. Due to limited representation of the very low-SES population in most research datasets, novel data linkages are necessary to understand how health and social policies impact this vulnerable population. Our results represent the first steps toward informing how integrating safety-net systems across diverse sectors (i.e., health, housing, and food) might enhance benefits to low-SES families. Our highlighted multisector data linkage, as part of the HCL Database, offers promise for future transdisciplinary researchers interested in the very low-SES population.

References

Adams, Peter, Michael. D. Hurd, Daniel L. McFadden, Angela Merrill, and Tiago Ribeiro. 2003. Healthy, wealthy, and wise? Tests for direct causal paths between health and socioeconomic status. *Journal of Econometrics* 112 (1): 3–56.

Adler, Nancy E., and Joan M. Ostrove. 1999. Socioeconomic status and health: What we know and what we don't. *Annals of the New York Academy of Sciences* 896 (1): 3–15.

Ahrentzen, Sherry. 2003. Double indemnity or double delight? The health consequences of shared housing and "doubling up." *Journal of Social Issues* 59 (3): 547–68.

Andersen, Virginia, Sarah Austin, Joel Doucette, Ann Drazkowski, and Scott Wood. 2015. Addressing income volatility of low income populations. Workshop in Public Affairs, Robert M. La Follette School of Public Affairs, University of Wisconsin–Madison.

Berkowitz, Michael K., and Jiaping Qiu. 2006. A further look at household portfolio choice and health status. *Journal of Banking & Finance* 30 (4): 1201–17.

Blackman, D. J., and C. M. Masi. 2006. Racial and ethnic disparities in breast cancer mortality: Are we doing enough to address the root causes? *Journal of Clinical Oncology* 24 (14): 2170–78.

Bradley, Cathy J., David Neumark, and Meryl Motika. 2012. The effects of health shocks on employment and health insurance: The role of employer-provided health insurance. *International Journal of Health Care Finance and Economics* 12 (4): 253–67.

Bradley, Cathy J., Lynne Penberthy, Kelly J. Devers, and Debra J. Holden. 2010. Health services research and data linkages: Issues, methods, and directions for the future. *Health Services Research* 45 (5p2): 1468–88.

Condliffe, Simon, and Charles R. Link. 2008. The relationship between economic status and child health: Evidence from the United States. *The American Economic Review* 98 (4): 1605–18.

Conley, Dalton, and Jason Alan Thompson. 2011. Health shocks, insurance status and net worth: Intra-and inter-generational effects. National Bureau of Economic Research (NBER) Working Paper 16857, Cambridge, MA.

Currie, Janet. 2008. Healthy, wealthy, and wise: Socioeconomic status, poor health in childhood, and human capital development. NBER Working Paper 13987, Cambridge, MA.

Currie, Janet, and Douglas Almond. 2011. Human capital development before age five. *Handbook of Labor Economics* 4:1315–486.

Curtis, Marah A., Hope Corman, Kelly Noonan, and Nancy E. Reichman. 2013. Life shocks and homelessness. *Demography* 50 (6): 2227–53.

Deaton, Angus. 2008. Income, health and wellbeing around the world: Evidence from the Gallup World Poll. *The Journal of Economic Perspectives: A Journal of the American Economic Association* 22 (2): 53–72.

Doyle, Joseph J. Jr. 2005. Health insurance, treatment and outcomes: using auto accidents as health shocks. *Review of Economics and Statistics* 87 (2): 256–70.

Drewnowski, A., C. D. Rehm, and D. Solet. 2007. Disparities in obesity rates: Analysis by ZIP code area. *Social Science & Medicine* 65 (12): 2458–63.

Edin, Kathryn, Melody Boyd, James Mabli, Jim Ohls, Julie Worthington, Sara Greene, Nicholas Redel, and Swetha Sridharan. 2013. *SNAP food security in-depth interview study*. Nutrition Assistance Program Report Series. Alexandria, VA: Office of Research and Analysis, Food and Nutrition Service

ESRI, and Tele Atlas. 2012. ESRI StreetMap premium: North America. Redland, CA: ESRI.

García-Gómez, Pilar. 2011. Institutions, health shocks and labour market outcomes across Europe. *Journal of Health Economics* 30 (1): 200–13.

Heckman, James J., and Jeffrey A. Smith. 1999. The pre-programme earnings dip and the determinants of participation in a social programme. Implications for simple programme evaluation strategies. *Economic Journal* 109 (457): 313–48.

Kim, Hyungsoo, Wonah Yoon, and Karen A. Zurlo. 2012. Health shocks, out-of-pocket medical expenses and consumer debt among middle-aged and older Americans. *Journal of Consumer Affairs* 46 (3): 357–80.

Lee, Jinkook, and Hyungsoo Kim. 2008. A longitudinal analysis of the impact of health shocks on the wealth of elders. *Journal of Population Economics* 21 (1): 217–30.

Marmot, M. G., S. Stansfeld, C. Patel, F. North, J. Head, I. White, E. Brunner, A. Feeney, and G. Davey Smith.1991. Health inequalities among British civil servants: The Whitehall II study. *The Lancet* 337 (8754): 1387–93.

Morduch, Jonathan, Timothy Odgen, and Rachel Schneider. 2014. *An invisible finance sector: How households use financial tools of their own making*. U.S. Financial Diaries Issue Briefs 3. Available from http://www.usfinancialdiaries.org/issue3-informal.

Murray, Alan T., Tony H. Grubesic, Ran Wei, and Elizabeth A. Mack. 2011. A hybrid geocoding methodology for spatio-temporal data. *Transactions in GIS* 15 (6): 795–809.

Noonan, Kelly, Hope Corman, and Nancy E. Reichman. 2014. Effects of maternal depression on family food insecurity. NBER Working Paper 20113, Cambridge, MA.

Parchman, Michael L., and Steven D. Culler. 1999. Preventable hospitalizations in primary care shortage areas: An analysis of vulnerable Medicare beneficiaries. *Archives of Family Medicine* 8 (6): 487.

Peek, M. E., A. Cargill, and E. S. Huang. 2007. Diabetes health disparities: A systematic review of health care interventions. *Medical Care Research and Review* 64 (5 suppl): 101S–156S.

Semyonov, Moshe, Noah Lewin-Epstein, and Dina Maskileyson. 2013. Where wealth matters more for health: The wealth-health gradient in 16 countries. *Social Science & Medicine* 81:10–17.

Smith, James P. 1998. Socioeconomic status and health. *American Economic Review* 88 (2): 192–96.

Smith, James P. 2004. Unraveling the SES: health connection. *Population and Development Review* 30:108–32.

Smith, James P. 2005. Consequences and predictors of new health events. In *Analyses in the Economics of Aging*, ed. David A. Wise, 213–40. Chicago, IL: University of Chicago Press.

Smith, James P., and Raynard Kington. 1997. Demographic and economic correlates of health in old age. *Demography* 34 (1): 159–70.

van der Loo, Mark P. J. 2014. The stringdist package for approximate string matching. *The R Journal* 6:111–22.

van Doorslaer, Eddy, Adam Wagstaff, Han Bleichrodt, Samuel Calonge, Ulf-G Gerdtham, Michael Gerfin, Jose Geurts, Lorna Gross, Unto Häkkinen, and Robert E Leu. 1997. Income-related inequalities in health: some international comparisons. *Journal of Health Economics* 16 (1): 93–112.

Wu, Stephen. 2003. The effects of health events on the economic status of married couples. *Journal of Human Resources* 38 (1): 219–30.

Weather-Related Hazards and Population Change: A Study of Hurricanes and Tropical Storms in the United States, 1980–2012

By
ELIZABETH FUSSELL,
SARA R. CURRAN,
MATTHEW D. DUNBAR,
MICHAEL A. BABB,
LUANNE THOMPSON,
and
JACQUELINE MEIJER-IRONS

Environmental determinists predict that people move away from places experiencing frequent weather hazards, yet some of these areas have rapidly growing populations. This analysis examines the relationship between weather events and population change in all U.S. counties that experienced hurricanes and tropical storms between 1980 and 2012. Our database allows for more generalizable conclusions by accounting for heterogeneity in current and past hurricane events and losses and past population trends. We find that hurricanes and tropical storms affect future population growth only in counties with growing, high-density populations, which are only 2 percent of all counties. In those counties, current year hurricane events and related losses suppress future population growth, although cumulative hurricane-related losses actually elevate population growth. Low-density counties and counties with stable or declining populations experience no effect of these weather events. Our analysis provides a methodologically informed explanation for contradictory findings in prior studies.

Keywords: population; migration; hurricanes; weather; disaster events; losses

Scientific warnings that climate-related extremes will increasingly impact ecosystems and social systems have focused scholarly attention on social and demographic responses to weather-related hazards, particularly the

Elizabeth Fussell is an associate professor of population studies (research) at Brown University. She studies population change in New Orleans after Hurricane Katrina as well as Latin American migration to the United States.

Sara R. Curran is a professor of sociology, international studies, and public policy & governance and the director of the Center for Studies in Demography and Ecology at University of Washington. She investigates internal migration in developing countries, globalization, family demography, environment and population, and gender.

Correspondence: elizabeth_fussell@brown.edu

DOI: 10.1177/0002716216682942

migration response (IPCC 2014). Many have interpreted this to mean that any population living in an area susceptible to an extreme climate event will out-migrate (e.g., Myers 2002). In reaction, a burgeoning interdisciplinary literature seeks to move beyond environmental determinism to consider how weather, geography, and society produce adaptive responses that include, but are not limited to, migration (cf. Black et al. 2011; Morss et al. 2011). But there are more hypotheses about this complex relationship than there are data to test them. Further, most research occurs within disciplines despite the interdisciplinary nature of the research, and population scientists are among those least engaged in research on environmental drivers of population change (de Sherbinin et al. 2007; Gall, Nguyen, and Cutter 2015; Hunter, Luna, and Norton 2015).

For decades, demographers focused their attention on how migration affects environmental outcomes, viewing migration as a key factor influencing environmental degradation (de Sherbinin et al. 2008; Hugo 2013). Recently, demographers reversed the causal order to ask how the environment drives migration (Hunter, Luna, and Norton 2015). They found heterogeneous effects of the environment on migration. For example, Gutmann and Field (2010) argue that catastrophes such as Hurricane Katrina or the 1930s Dust Bowl have relatively small impacts on population redistribution compared with the effects of environmental amenities (e.g., milder climates, proximity to water or mountains) or the management of environmental barriers (e.g., air conditioning, flood control, drainage, and irrigation). One example of migration to attractive but risky places is found in the growth of the population in coastal areas despite hurricane-related losses along the Atlantic hurricane coast (NOAA 2013). This is exactly the opposite of environmental determinists' prediction that people will move away from hazards.

Earlier demographers argued for modeling a dynamic relationship between population and environment, but data to do so were lacking (Bilsborrow 1992; Davis and Bernstam 1991). Our approach returns to this earlier articulation by building a dataset with sufficient temporal and spatial variability to model a reciprocal relationship between environment and population. To get at this dynamic relationship, we examine how past population trends, population density, current and cumulative weather events, and weather-related losses intersect at the county level to influence future population change. We hypothesize that the

---

Matthew D. Dunbar is assistant director of the Center for Studies in Demography and Ecology and affiliate assistant professor in geography at University of Washington. His research spans the field of spatial demography.

Michael A. Babb is a geospatial research scientist at the Center for Studies in Demography and Ecology. His research is on internal migration and racial structures in the United States.

LuAnne Thompson is the Walters Professor in the School of Oceanography and director of the Program on Climate Change at the University of Washington. Her research concerns the role that the oceans play in climate variability.

Jacqueline Meijer-Irons is a demographic research scientist at the Center for Studies in Demography and Ecology at University of Washington. Her research focuses on individual and household-level responses to environmental stress in rural communities in Thailand.

effect of weather hazards on future population growth is small compared with the effect of past population trends.

Weather Hazards and Population Change

Early research on weather hazards and population change in the United States concluded that there was no population impact of weather-related hazards on average (Wright et al. 1979). Since Hurricane Katrina struck the Gulf Coast in 2005, the question has been revisited, but findings are contradictory. For example, Pais and Elliott (2008) show that the four "billion dollar storms" of the early 1990s were associated with postevent population growth. Moving from hurricane-related losses to all hazard losses, Schultz and Elliott (2013) show that a county experiencing $1 million in total property damage from hazards during the 1990s grew in population by an average of 3.2 percentage points more than a county experiencing no disaster-related property damage between 1995 and 2000. Pais and Elliott (2008) speculate that the repair of hazard damage and opportunities to improve disaster-affected areas create jobs and housing that attract new residents. This line of research indicates that hazards produce a churning of the population, with some sociodemographic groups increasing and others decreasing but typically resulting in a net population gain (Elliott 2014).

Other research shows that natural hazard exposure is associated with reduced growth. For example, Logan, Issar, and Xu (2016) found that between 1970 and 2005, Gulf Coast counties with higher levels of hurricane-related damage experienced reduced growth for up to three years after the hurricane event, especially in counties with low poverty rates. Similarly, Shumway, Otterstrom, and Glavac (2014) used county-to-county migration flow data from 2000 to 2010 to show that counties with high environmental hazard impacts lose residents through net out-migration, and those out-migrants tend to be higher income residents who move to counties with lower cumulative hazard losses. These studies provide evidence that damage-related losses from hurricanes and other natural hazards suppress population growth when those who can afford to move to less hazardous places choose to do so and they are not replaced by in-migrants.

We suspect that this apparent contradiction in the research on the population effects of hurricanes is due to long-term trends in population growth, which influence how residents of disaster-affected areas respond. We know of no research addressing this question explicitly. There is, however, one study that found that ninety-two U.S. communities that experienced tornados in which at least half of their structures suffered major damage between 1992 and 2008 and that were already experiencing declining populations before the disaster were more likely to experience large postdisaster population losses (Cross 2014). Much like case studies that focus on catastrophic events, this study selected only those places that suffered extensive damage to structures, limiting the generalizability of the conclusions.

By examining all counties affected by hazards and incorporating long-term population trends and hazard events and losses, we test the hypothesis that the effect of weather hazards on future population growth is small compared with the effect of past population trends. While we expect that our hypothesis will be supported, evidence inconsistent with our hypothesis may help to reconcile this contradiction in the existing research on how populations change after an environmental hazard event and which types of places are most likely to experience population change.

Data and Methods

Data

To obtain the spatial and temporal variability in population trends and weather hazard events and losses necessary to accomplish our research objectives, we integrated county-level annual population estimates from the U.S. Census Bureau (2016) with the Spatial Hazard Events and Losses Database for the United States (SHELDUS; HVRI 2015). For the measures of population change, we included annual, county-level estimates from 1970 through 2012. Intercensal estimates were used for the 1990s and 2000s, while postcensal techniques were used for the 1970s, 1980s, and 2010s. We did not include intercensal estimates for the 1960s because they were produced with methods that are inconsistent with later estimates. Our population change dataset treats county-year as the unit of analysis and measures annual population size for each county-year.

The SHELDUS dataset measures annual county-level fatalities, injuries, and property and crop losses associated with eighteen types of natural hazard events in the United States from January 1960 to December 2014. SHELDUS combines data from twenty-three sources, though most come from the National Centers for Environmental Information data products. The loss estimates are obtained from emergency managers, U.S. Geological Survey, U.S. Army Corps of Engineers, power utility companies, and newspaper articles. These amounts refer to losses associated with damage to private property, including structures, objects, and vegetation, as well as public infrastructure and facilities. Damages or loss amounts are distributed evenly between counties in a multicounty event. As in the population data, the unit of analysis is a county-year. The population and hazard event data files were merged using ArcGIS geo-referenced county-year FIPS codes and county boundary files to produce a spatial-temporal database of county-years for each hazard type. We adjusted county boundaries to 2010 boundaries.[1] From the SHELDUS database, we selected only weather-related hazards, which include avalanches, coastal storms, droughts, floods, fog, hail storms, heat waves, hurricanes/tropical storms, landslides, lightning storms, severe thunderstorms, tornados, wind, and winter weather. We focused on these because we expect to see more of these types of events as climate change progresses. This excludes wildfires, earthquakes, tsunamis/seiches, and volcanic eruptions, none of which are directly attributable to climate change: wildfires are typically started by human activity and geologic activity is not climate-related.

Our database provides an important corrective to previous approaches to how populations respond to hazard impacts. Analyses that focus on population impacts of a single hazard event in a specific place commit two errors that threaten the generalizability of their findings: (1) they select only the most damaging and costly events and thereby neglect the full range of events, and (2) they ignore the cumulative impacts of previous hazards, a source of unobserved heterogeneity. By using data from a long period of time and for the entire affected region and that include all weather hazard events, we address both concerns.

We addressed these sources of unobserved heterogeneity by measuring all hazard events and placing the effect of a single hazard event in the context of past hazard events, hazard-related losses, and past population growth. We observed hazard events and losses for the years 1970 to 2009 and population data from 1970 to 2012. We then constructed decadal measures of cumulative hazard events and losses. A decade is long enough to remove much of the random element of hazard occurrence and capture secular trends. The measures of cumulative hazard events and hazard losses sum those annual quantities over the previous 10 years. The past population trend is defined by a 10-year compound average population growth rate: $CAPGR = \left(\dfrac{pop_t}{pop_{t-10}}\right)^{\left(\frac{1}{11}\right)} - 1$, where pop is the county population in a given county-year, and t is the reference year. Therefore, our analysis begins in 1980, the first year in which we have 10 years of past hazard and population data. We measure future population growth as a three-year compound average annual population growth rate: $CAPGR = \left(\dfrac{pop_{t+3}}{pop_t}\right)^{\left(\frac{1}{4}\right)} - 1$, where pop is the county population in a county-year, t, and t is the reference year. We chose three years as a time frame because it allows for the possibility of population recovery and growth (or loss) after a hazard event. We also included a measure of county population density, defined as the population per square mile. Based on our decision to measure future growth in three-year intervals and data availability, we end our analysis in 2012.

Our treatment of hazard losses departs from much of the social science research on this topic. Most research tends to treat hazard losses as equivalent regardless of the hazard type producing the loss. We propose that losses are not equivalent; for example, damage to a home resulting in temporary or permanent loss of its use is different than damage to utility infrastructure that causes loss of access to electricity, gas, or water until services are restored. While the former displaces residents for prolonged periods, the latter does not. Differences in types of damage produced by hazard events means that they are unequally likely to produce population change even when monetized loss estimates are equivalent. Therefore, we consider hazards by type, and here we focus on hurricanes and tropical storms (hereafter, hurricanes) because they are among the most damaging weather events and are more likely to produce a migration response.

Methods

Our hypothesis is that the effect of a weather hazard in a given year on future population growth is small compared with the cumulative effects of weather hazards and past population trends. To test this, we estimate a random effects linear regression that takes a reduced form, evaluating the marginal impacts of hazards on future population change and controlling for past population trends. We assume that while political, social, and economic conditions and trends also influence future population growth, their influence is gauged by past population growth. Past population growth captures a large portion of this unobserved heterogeneity. To further address the issue of unobserved heterogeneity between counties, we analyze our county-year data with a random effects generalized least squares regression estimation model (STATA xtreg re):

$$y_{it} = x'_{it}(\beta_{it} + h_i) + (\alpha_{it} + u_i) + \varepsilon_{it}$$

This random effects estimator is a matrix-weighted average of the fixed-effects (within) and the between effects, where y_{it} is the outcome, the prospective three-year compound annual growth rate from time t, for every county i in year t; x'_{it} is a vector of county-year factors that vary across time and counties; and β_{it} is the between-effects parameter.[2] The coefficients are interpreted as the average effect of a county-year factor on the future population growth rate, *ceteris paribus*.[3] Since the data come from all U.S. counties that have experienced hurricanes between 1970 and 2012, our data constitute the entire population (not a sample) of county-years, and we report tests of statistical significance as an indication of the meaningfulness of the estimated coefficients (e.g., how different they are from zero).

In the analysis section, we explore the spatial and temporal variability of hazard-related losses to justify our focus on single hazard impacts, in this case, hurricanes. We follow this with a brief description of the spatial and temporal variability in population trends and provide a justification for evaluating our hypotheses on different county-year data subsets. Next, we discuss the descriptive statistics for our dataset. We then present the results of our regression analysis to test our hypothesis that the effect of a hurricane in a given year on future population growth is small compared with the cumulative effects of weather hazards and past population trends. In the final section, we summarize our findings and their implications for future research on weather-related hazards and population change.

Analysis

Spatial and temporal variability in hazard events

To illustrate the spatial and temporal variability in hazardous weather events, we use 2008 as our reference county-year, since this was a year in which total property and crop losses due to hazards was high ($25.3 billion),[4] but not extraordinary as in 2005 ($122.3) or 1992 ($56.1), when especially destructive hurricanes impacted large metropolitan areas.[5] In 2008, spatial variability in hazard-related combined losses were mostly due to spring flooding in the Midwest and

FIGURE 1
Total Property and Crop Losses Due to All Weather-Related Hazards in 2008 (A) and the Previous Decade (1998–2007; B), Both Adjusted to 2014 Dollars

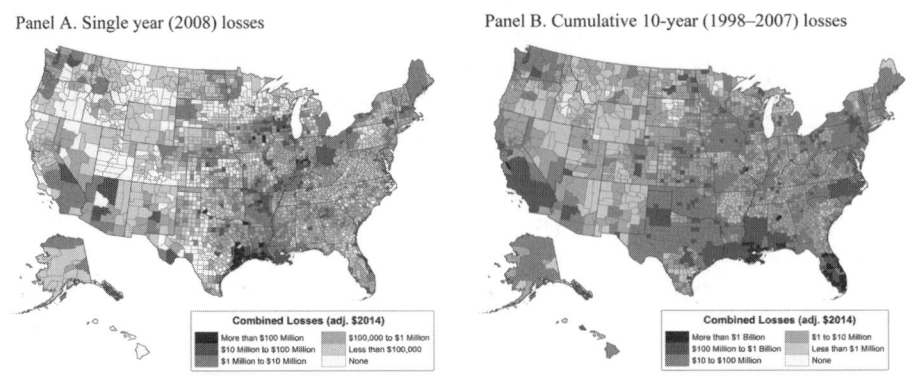

Hurricanes Ike and Gustav, which struck the Texas and Louisiana coasts and traveled inland (see Figure 1, Panel A). However, a current-year measure of hazard impact neglects the spatial variability in cumulative hazard losses over the prior decade (1998 to 2007), which is quite different (see Figure 1, Panel B). Observing cumulative hazard losses captures a spatial and temporal dimension of heterogeneity—the spatially unequal accumulation of hazard events and losses—unobserved in earlier studies that tends to focus on the effect of single events. Based on this observation, we expect that population responses to hazard events and losses in a single year will be different than population responses to the cumulative number of events and amount of losses.

Prior research often uses total hazard-related losses to measure hazard impacts. However, by aggregating all hazards that have unequal probabilities of occurring across geography and are unequally likely to produce the same type of damage, total hazard losses are a noisy predictor of population change. For example, Figure 2 shows total cumulative combined losses due to hazards (Panel A), which amounted to $266.2 billion (held constant to US$2014) over 1998 to 2007, and hazard-specific cumulative combined losses in the other five panels. In each county, specific hazards account for different proportions of cumulative losses. Hurricanes, which account for nearly two-thirds of cumulative losses nationally, produce the greatest losses in coastal areas of the Gulf of Mexico and the Eastern seaboard, causing severe damage to private residential and business properties as well as built infrastructure and the environment (Panel B). Flooding losses—9.5 percent of cumulative losses—concentrate in parts of the Southwest, the South, the Midwest, and the Northeast and cause similar damages as hurricanes (which they often accompany) but over a smaller geographic area (Panel C). Losses from drought—6.5 percent—are concentrated in agricultural regions and principally destroy crops (Panel D). Tornado-related losses—4.9 percent—are spatially focused and dispersed through "tornado alley" in the Midwest and the Eastern seaboard, causing severe destruction over very defined areas (Panel E).

FIGURE 2
Total Cumulative Property and Crop Losses in a Decade (1998–2007) Due to All Weather-Related Hazards (A) and the Five Costliest Hazards (B–F)

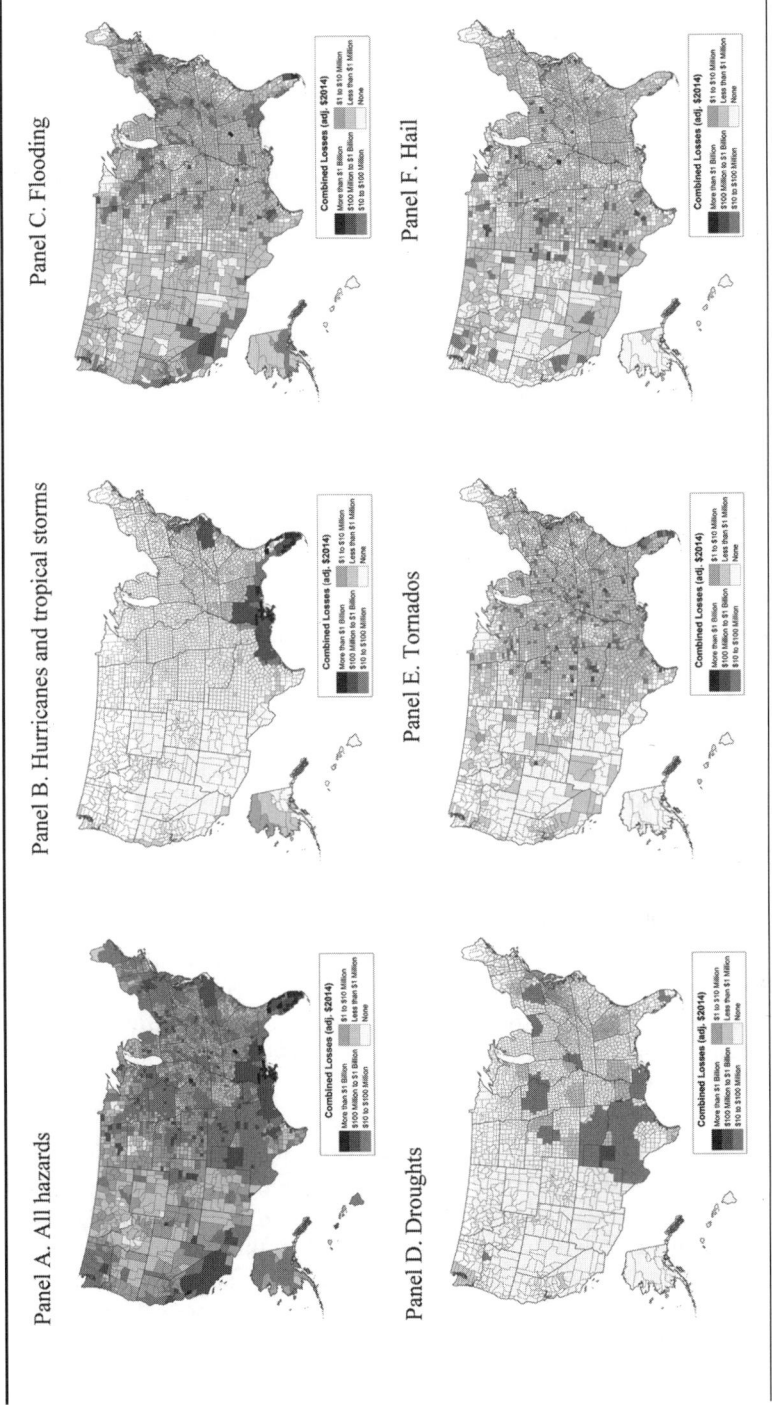

FIGURE 3
Comparison of County-Level Hurricane and Tropical Storm Property and Crop Losses (A) and Population Change (B) from 1970 to 2009

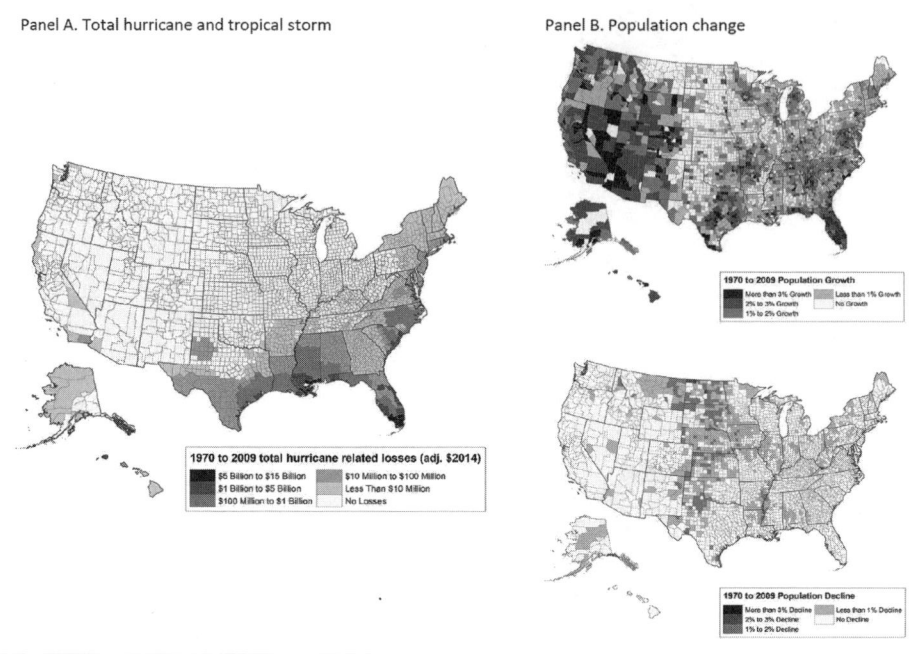

Hail-related losses—4.4 percent—are spatially focused and widely dispersed, and may damage property but not as severely as the costlier hazards listed above (Panel F). Considering this variability in hazard geography and given our inference that associated losses are qualitatively different among hazard types, we chose to focus on total property and crop losses from one hazard type, hurricanes, which are spatially limited and costly.

Between 1970 and 2009, the years for which our data allow us to investigate hazards and population change in this analysis, virtually all of the hurricane-related losses occurred in the Gulf of Mexico and the Atlantic Coast. A few Pacific storms produced limited damage-related losses on the West Coast, Alaska, and Hawaii. By pairing the maps of cumulative hurricane-related losses with the maps of population change (growth and decline) over the same period, it is evident that many counties in Texas, Louisiana, Florida, and the Carolinas with the greatest cumulative losses also experienced sizable population growth (see Figure 3). This spatial correlation reflects two things. First, more densely populated counties have more assets exposed to hazards and therefore experience greater losses. Second, and important for our study, these areas continue to increase in population. On the surface, this would suggest that hurricanes do not discourage human settlement.

FIGURE 4
Variability in Population Density (A), Historic Population Trends (B), and Future Growth for Counties in Reference Year 2008 (C)

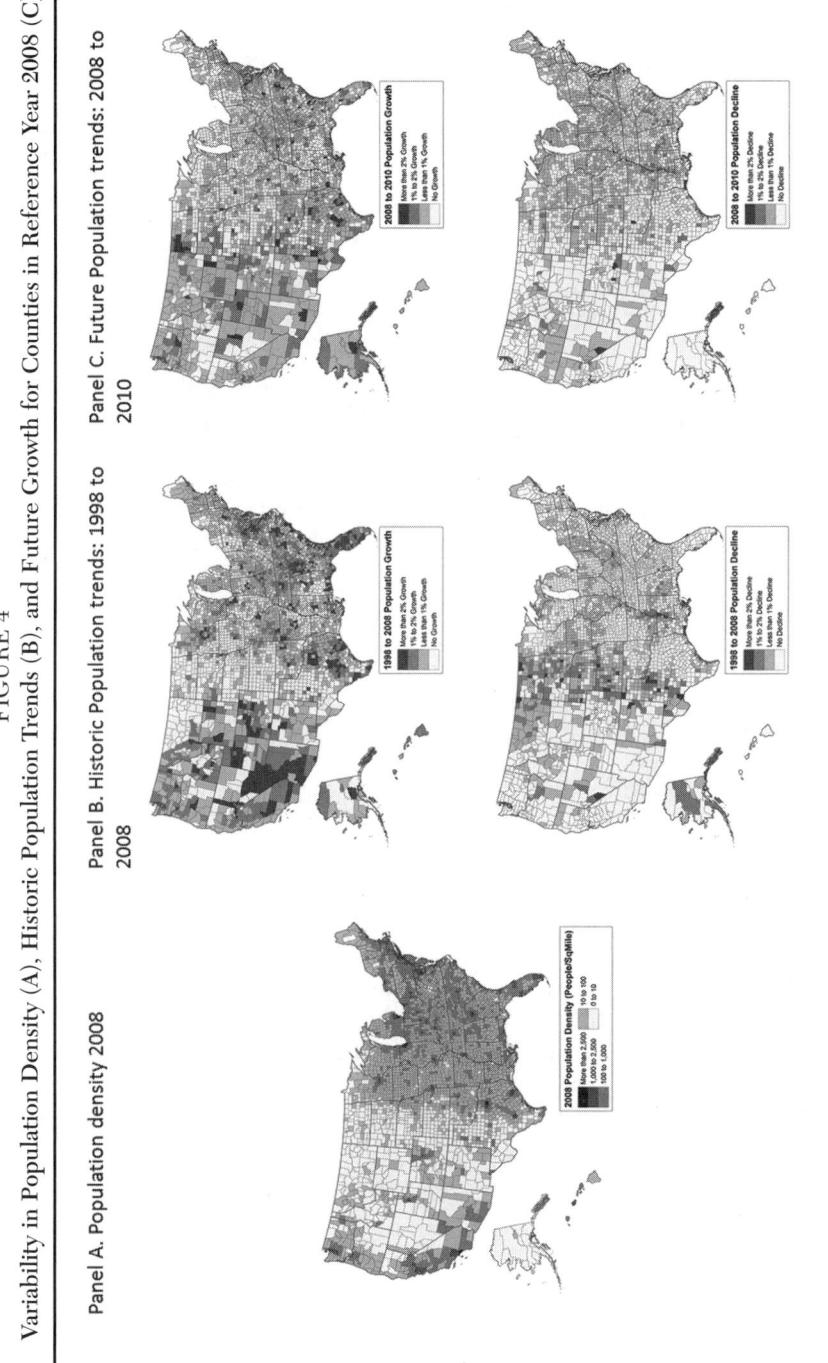

Spatial and temporal variability in population density and growth

To illustrate the spatial distributions of our population measures, we show population density, past population growth, and future population growth using 2008 as a reference year (see Figure 4). Population density (Panel A) shows densities are higher on average in the eastern United States and the West Coast compared with counties ranging from the western mountains through the arid Midwest. Panel B shows past population trends for county-year 2008, referencing the prior decade, 1998–2007. During this period, counties in the East and West grew the fastest, while many counties throughout the middle of the country declined, a few by more than 2 percent. Overall, most counties evince little change, registering less than 2 percent growth or decline. Panel C shows future population growth for county-year 2008, referencing change from 2008 to 2010. This shows that prospective population growth rates across U.S. counties are mostly inclining. Even for counties with declining past population growth rates, the forward projection appears to be one of little to slow growth with very few counties showing rates of decline greater than 2 percent.

Next, we complement our discussion of spatial variability in hazards and population with a discussion of temporal variability in population trends. We find that while past population growth rates and future population growth are relatively stable, variance around these annual means has been decreasing with each year. In contrast, means for population density have also been stable, but variance has been increasing with each year (analysis not shown). The mixed results of earlier studies may be due to these changes in variability with time and across space, since results will vary depending on the counties chosen and the time frame evaluated.

Delving further into our investigation of differences between counties over time, we examined differences in future population growth between high- and low-density counties and counties with past inclining and declining growth rates (analysis not shown). We found that counties with declining past population trends, especially low-density counties, tend to have higher rates of future growth, especially when past declines were large. In contrast, counties with inclining past population trends, especially high-density counties, tend to have higher rates of future growth, especially when past inclines were large. Because these underlying population trends are so different, an analysis of the complete spatial-temporal database is likely to obscure the effects of hurricanes. To more cleanly expose population trends after a hurricane, we subset our data into four categories of counties: high (density $\geq 1,000$) and low (density $< 1,000$) density and inclining (CAPGR > 0) and declining (CAPGR ≤ 0) past growth trends. We use these subsets in our multivariate random effects regression analysis in the next section.

Descriptive statistics for the spatial-temporal database and subsets

Distinguishing counties by past growth trends (CAPGR) and population density is a simple way to discern the heterogeneity of current year and cumulative

TABLE 1
Population Trends and Hazard Exposures for All Counties Ever Experiencing a Hurricane

All County-Years 1980–2009, among Counties Ever Exposed to Hurricanes between 1970–2009; means (SD)

	All Counties	Counties with Declining 10-Year Population Trend		Counties with Inclining 10-Year Population Trend	
		Pop density < 1,000 people/ square mile	Pop density ≥ 1,000 people/ square mile	Pop density < 1,000 people/ square mile	Pop density ≥ 1,000 people/ square mile
N	39,314	28,999	2,010	7,481	824
Population variables					
Compound annual pop growth rate from year = t to $t + 3$	.0061 (0.019)	.00821 (.012)	.007 (.008)	-.002 (.010)	-.001 (.012)
Compound annual pop growth rate from year = $t - 11$ to t	-.00898 (0.0129)	-.01307 (.012)	-.01 (.009)	.005 (.006)	.005 (.006)
Population density (people per square mile) in year t	418.785 (2503.25)	129.085 (172.023)	4156.024 (8907.199)	71.272 (137.926)	4652.892 (6951.653)
Hazard variables (hurricanes)					
$ losses (million)/capita (USD 2014) in past decade	.000981 (0.008)	.00085 (.005)	.0001 (.0001)	.002 (.014)	.0003 (.003)
$ losses (million)/capita (USD 2014) in a county-year	.00012 (.002)	.00012 (.002)	.00001 (.0002)	.0001 (.002)	.00004 (.001)
# of hurricanes in past decade	1.272 (1.979)	1.296 (2.005)	1.326 (2.007)	1.111 (1.768)	1.742 (2.601)
Any hurricane in a county-year	.091 (.287)	.094 (.292)	.105 (.307)	.073 (.260)	.101 (.301)

hurricane events and losses (see Table 1). The vast majority (93 percent, or 36,480 of 39,314) of county-years were for counties with low population densities, and, of these, 80 percent (28,999 of 36,480) had declining past growth trends. Of the remaining county-years for counties with high population densities (7 percent, or 2,834 of 39,314), the majority (71 percent, or 2,010 of 2,834) had declining past growth. Only 2 percent (824 of 39,314) of all county-years were contributed by high-density counties with inclining past growth trends. These unequal group sizes mean the low-density, declining growth counties have an undue weight in any regression analysis, which is why we treat them separately in the regression analysis.

Counties with declining 10-year growth trends tend to experience growth in the next three years, regardless of population density. Counties with inclining 10-year growth trends tend to experience population loss in the next three years, regardless of population density. There are fewer differences between county-types in hurricane events and losses in both current year and the past decade. Counties that ever experienced a hurricane between 1970 and 2009 experienced at least one hurricane on average in 9.1 out of every hundred county-years, with large variability between counties (mean = 0.091; SD = 0.287). The total amount of property and crop losses in the current year adjusted to 2014 U.S. dollars and presented as millions per capita provides a measure of losses relative to the number of people at risk.[6] In the average county-year, these losses amount to $120 per capita (mean = 0.00012; SD = 0.002). In general, per capita losses are greater in low-density than high-density county-years. The accumulated hurricane-related losses are the total amount of property and crop losses summed for each of the 10 years prior to the current year adjusted to 2014 U.S. dollars and presented as millions per capita. In an average county-year, the cumulative losses amount to $981 per capita (mean = 0.000981; SD = 0.008). The cumulative count of hurricanes is a sum of hurricane and tropical storm for the previous 10 years. In an average county-year, the county had experienced 1.272 (SD = 1.979) hurricanes in the previous 10 years. We use these measures in our multivariate analysis to examine differences in hurricane effects between county-years.

Multivariate random effects regression analysis

To test our hypothesis that the effect of hurricanes on future population growth is small compared with past population trends, we estimate regression models for each subset of the data. In Table 2, we focus on low- and high-density county-years with declining population growth trends, and in Table 3, we focus on low- and high-density county-years with inclining population growth trends. Each panel of the two tables includes three models: model 1 includes only the population variables, past CAPGR, and population density; model 2 includes the four hurricane variables, current year and past decade hazard losses and events; and model 3 includes both the population and hazard variables. This allows us to evaluate the relative contribution of each set of variables to explaining variance in future population growth. We fully interpret the analysis of county-years with declining past population trends and low population densities (see Table 2, Panel A), and summarize findings for the remaining panels in Tables 2 and 3.

TABLE 2
Random Effects Linear Regression for Three-Year Prospective Population Growth Rate among Counties with Declining Population Growth Trends (County-Years 1980–2009 Ever Exposed to Hurricanes Between 1970–2009); coefficient (SE)

PANEL A (< 1,000 pop/square mile)	Model 1	Model 2	Model 3
Historic compound annual pop growth rate	−.291 (.007)°°°		−.285 (.007) °°°
Population density	−3.90E–06 (7.20E–07)°°°		−3.60E–06 (7.10E–07)°°°
$ losses (million)/capita (USD 2014) in past decade		−.079 (.013) °°°	−.073 (.012) °°°
$ losses (million)/capita (USD 2014) in county-year		−.201 (.019) °°°	−.213 (.021) °°°
# of hurricanes in past decade		−.0005 (.00004) °°°	−.0003 (.00004)°°°
Any hurricane in a county-year		.0006 (.0002) °°°	.0009 (.0002) °°°
Constant	.004 (.0002)°°°	.007 (.0003) °°°	.004 (.0002) °°°
R-square within (R-square between)	.013 (.655)	.016 (.009)	.022 (.646)
PANEL B (≥ 1000 pop/square mile)	**Model 1**	**Model 3**	**Model 4**
Historic compound annual pop growth rate	−.272 (.024) °°°		−.265 (.024) °°°
Population density	−4.65E–08 (4.46E–08)		−4.80E–08 (4.38E–08)
$ losses (million)/capita (USD 2014) in past decade		1.028 (.322) °°°	.884 (.301) °°
$ losses (million)/capita (USD 2014) in county-year		.829 (.701)	.802 (.702)
# of hurricanes in past decade		−.0004 (.0001)°°°	−.0002 (.0001)
Any hurricane in a county-year		−.0001 (.0004)	−.0001 (.0004)
Constant	.004 (.001)°°°	.007 (.0006) °°°	.004 (.0005)°°°
R-square within (R-square between)	.017 (.623)	.014 (.0006)	.0234 (.6204)

°$p < .01$. °°$p < .005$. °°°$p < .001$.

TABLE 3
Random Effects Linear Regression for Three-Year Prospective Population Growth Rate among Counties with Inclining Population Growth Trends (County-Years 1980–2009 Ever Exposed to Hurricanes Between 1970–2009); coefficient (SE)

PANEL A (< 1,000 pop/square mile)	Model 1	Model 2	Model 3
Historic compound annual pop growth rate	.270 (.022) °°°		.214 (.023) °°°
Population density	2.72E–06 (2.03E–06)		3.41E–06 (1.74E–06)
$ losses (million)/capita (USD 2014) in past decade		.08 (.009) °°°	.055 (.001) °°°
$ losses (million)/capita (USD 2014) in county-year		–.022 (.045)	–.033 (.045)
# of hurricanes in past decade		–.0002 (.00009) °°	–.0002 (.00009)
Any hurricane in a county-year		–.001 (.0004) °	–.001 (.0004) °°
Constant	.0008 (.0004)	.0007 (.0003)	–.0006 (.0003)
R-square within (R-square between)	.027 (.009)	.0007 (.0598)	.0287 (.0003)

PANEL B (≥ 1,000 pop/square mile)	Model 1	Model 3	Model 4
Historic compound annual pop growth rate	.816 (.077)°°°		.312 (.090) °°°
Population density	8.66E–09 (7.32E–08)		4.71E–08 (6.92E–08)
$ losses (million)/capita (USD 2014) in past decade		2.953 (.218) °°°	2.463 (.253) °°°
$ losses (million)/capita (USD 2014) in county-year		–2.324 (.465) °°°	–2.379 (.464) °°°
# of hurricanes in past decade		–.0019 (.0003) °°°	–.0017 (.0003) °°°
Any hurricane in a county-year		–.0023 (.0012)	–.0022 (.001)
Constant	–.003 (.001)°°°	.003 (.0008) °°°	.001 (.0009)
R-square within (R-square between)	.206 (.196)	.289 (.0062)	.318 (.0053)

°$p < .01$. °°$p < .005$. °°°$p < .001$.

Among county-years in which past growth is negative and population density is low, we confirm the fluctuation in growth rates that we discerned in our investigation of means and variance. In model 1, the coefficient for past population growth has a negative value (–0.291). Since all the values of this variable in this data subset are negative by design, this coefficient is always multiplied by a negative value, so that a 1-unit decline in the past population growth rate yields a 0.291-unit increase in the future population growth rate. In other words, on average, a county with a pattern of past population decline experiences a lower future rate of decline, or possibly slightly positive growth. This interpretation applies only to this variable in county-years with declining populations (see Table 2). Model 1 also shows that population density has a trivial negative effect on future population growth: each additional person per square mile reduces the future population growth rate by 0.000004 units in counties with population densities less than 1,000 people per square mile. These two population measures explain about two-thirds of the variability in future population growth between counties (R-square = 0.655) but very little of the within-county variability in future population growth (R-square = 0.013).

Models 2 and 3 of Table 2 show the effects of hurricane losses and events on future growth, on their own and in combination with the population variables, and formally test our hypothesis. In model 2, the effects of current year and accumulated hurricane losses are both negative, as expected, and small. Each additional $1 million per capita loss in a particular year reduces the future growth rate by 0.2 units. The effect of each additional $1 million per capita in losses accumulated over 10 years reduces the population growth rate by 0.079 units. Given that hurricane loss amounts per capita are far below $1 million in any county-year, these impacts are small. Counties that experience at least one hurricane in a year experience a very slight (b = 0.0006) incline in the future population growth rate. However, this may be negated if there were recent hurricanes, since each additional hurricane event in the past decade is associated with a decline in the population growth rate (b = –0.0005). Consistent with our expectation that population trends are more important than hurricanes, hurricane events and losses do not explain much variability in future population growth, either between (R-square = 0.009) or within (R-square = 0.016) these lower density counties with declining rates of population growth.

The coefficients are relatively unchanged in model 3, which includes both population and hurricane variables. Model fit improves negligibly in model 3 relative to model 1 but is an improvement over model 2. Overall, population trends in low-density counties with declining growth rates are more important in predicting future population growth than hurricane events and losses. This confirms our hypothesis for this set of county-years.

For Table 2, panel B, high-density counties with declining population trends, the patterns are very similar to low-density counties with declining population trends (Panel A). Therefore, we focus only on model 3 in panel B. This model indicates that high-density counties undergoing past population declines experience slowing rates of population decline (b = –0.265) and population density exerts a trivial effect (b = 0.00000005) on future growth. Unlike counties in Panel

A, current year hurricane-related losses ($b = 0.80$) and accumulated hurricane losses ($b = 0.88$) positively impact future population growth. However, given that hurricane loss amounts per capita are far below $1 million in any county-year, these impacts are quite small. Like counties in panel A, current year ($b = -0.0001$) and cumulative ($b = -0.0004$) hurricane events exert very small negative effects on future population growth. The model fit statistics are also similar, indicating that population differences between these high-density counties with declining populations are more important in predicting future population growth than hurricane events and losses, again confirming our hypothesis for this set of county-years.

Turning to Table 3, in which we estimate models for low- and high-density counties with past inclining growth rates, we again focus on model 3 coefficients. In panel A, with model estimates for low-density counties, past population growth leads to increased ($b = 0.214$) future population growth, while population density has a trivial effect on future population growth. Hurricane losses have minor impacts on future growth: hurricane-related losses in a given year slightly decrease ($b = -0.033$) future population growth, and losses over the previous decade slightly increase ($b = 0.055$) future population growth. Hurricane events also have minor effects: county-years in which at least one hurricane struck experience a 0.001-unit decrease in future population growth rates, and each additional hurricane over the previous decade produces a very slight ($b = -0.0002$) reduction in future population growth. None of the models is particularly powerful in explaining the variance in population growth, either within or between county-years. In other words, low-density counties with inclining population growth rates are growing in response to factors other than past population trends and hurricane events and losses.

In contrast, Table 3, panel B, model 3, explains more of the variation in future population growth for high-density counties with inclining population growth trends. For these counties, higher past population growth rates yield positive effects ($b = 0.312$) on future population growth rates, although population density impacts ($b = 0.000003$) are trivial. As in low-density counties (Panel A), hurricane losses suffered in a particular year negatively affect future population growth, but the effect is larger: for each one million dollars of losses per capita in a given year, the future population growth rate is reduced by −2.38 units. However, higher amounts of per capita loss accumulated over the previous decade compensate for this and yield a 2.46-unit higher future population growth rate. The effects of a hurricane strike in the current year ($b = -0.0022$) and in the past decade ($b = -0.0017$) both tend to suppress future population growth. Notably, the coefficient for past population growth declines by 61 percent between model 1 and model 3, indicating that hurricane events and losses dampen the effect of past growth on future growth. In other words, unlike what we hypothesized, the effect of past growth on future growth depends to a degree on both current and cumulative hurricane events and losses. Specifically, current and cumulative hurricane events and current year losses tend to suppress future growth, while cumulative losses tend to increase it. The model fit statistics suggest that both population and hurricanes explain within-county variance, although

FIGURE 5
Prediction of Future Population Growth Given Covariate Values Based on Results from Regression Models

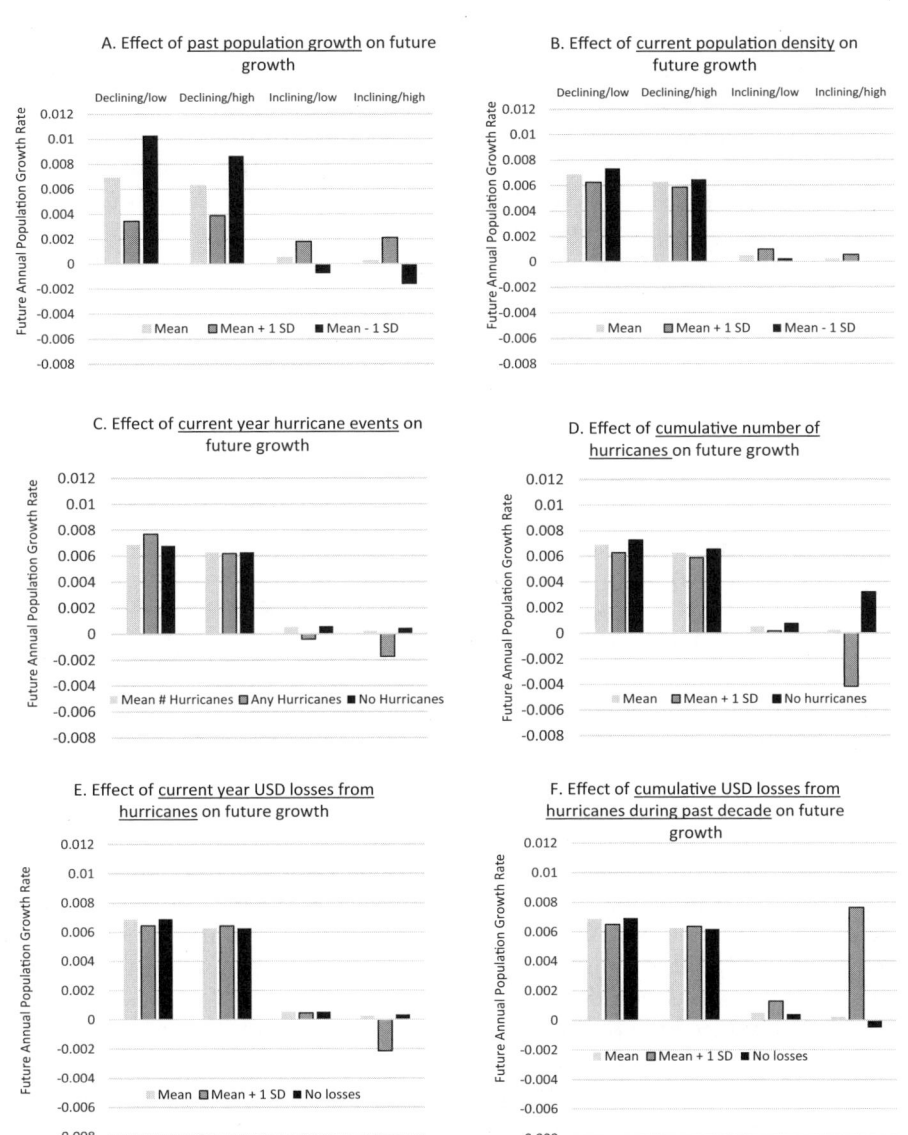

neither contributes much to between-county variance in the full model. From this set of models for all four county-year types, we can conclude that it is mainly these higher density counties with inclining population growth rates that experience population impacts of hurricane events and losses.

The findings of our regression analysis are more clearly displayed in Figure 5, which shows predicted future growth rates for each subset of county-years for each of the six variables of interest while holding all other variables at their mean. The six panels include four groups (county-year subsets) of three bars: the first bar is the predicted future growth with the variable of interest held at its mean value, and the second and third, respectively, are the predicted future growth with the variable held at the mean value plus or minus one standard deviation, or in the cases where the mean value minus one standard deviation falls into the negative range, we substitute zero. When the difference in height between bars is small for a subset of county-years, the effect of the variable on future growth is small; when the difference in bar heights is large, the effect is more meaningful.

Our hypothesis that the effects of hurricanes on future population growth is small compared with that of past population growth is clearly supported because the size of the effect of past population growth on future population growth is greater than the effect of any of the other variables shown here, with only a few notable exceptions (Figure 5). Panel A shows that within each subset of counties, the effects of past population growth on future population growth is sizable, especially in comparison to the effects of the hurricane variables (Panels C through F). The effects of past growth are especially large in the subsets of county-years with declining population trends: the greater the rate of past decline, the greater the rate of future growth. More modest effects are evident for the subsets of county-years with inclining past growth, where higher than average past growth is associated with higher future growth, on average. Population density (Panel B) shows small effects on future growth, which we do not discuss because the effects are statistically insignificant.

Among the hurricane-related variables, both current year and cumulative hurricane events and losses have relatively small effects on future growth in most county-year subsets (Figure 5, Panels C through F). These effects are trivial for county-years with declining population trends, regardless of population density, as is evident from the similarity in bar heights. Differences are also small and close to zero for county-years with inclining growth and low population density.

The notable exception to the pattern of small effects of hurricanes relative to past population trends is found for county-years with inclining past population trends and high population densities. These county-years show sizable effects of hurricanes on future growth (the last set of bars). For these county-years, future growth is suppressed by a hurricane in the current year (Panel C), a higher than mean number of hurricanes in the past decade (Panel D), and higher current-year hurricane-related losses (Panel E). However, higher cumulative losses over the past decade strongly increase future population growth (Panel F). This suggests that there is a dynamic relationship between cumulative hurricane losses and population growth in these high-density, growing counties that merits further investigation.

Conclusion

The existing demographic literature on weather hazards and population change in the United States is sparse and unsystematic. To analyze this relationship, we constructed a spatial-temporal database of U.S. counties from 1980 to 2012 with measures of population, weather events, and weather-related losses. We explored this spatial-temporal database to refine our approach to this problem. We concluded that the best way to model the problem is separately by hazard, recognizing that weather events are variable in their spatial and temporal distribution, mode of impact, and the value of associated losses. We focused on hurricanes and tropical storms because of their destructive power and their potential to displace people from their homes and communities. We differentiated between long- and short-term effects to estimate more precisely how hurricanes influence future population trends. We differentiated the counties according to past population growth trends and population density, and used a reduced form, random effects linear regression model to test our hypothesis that the effects of weather hazards on future population growth are very small compared with the effect of past population trends.

We find support for our hypothesis in counties with past population declines regardless of their population density. These are the majority of hurricane-affected U.S. counties, and in these places, hurricanes hardly affect future population; past population trends are far more predictive of future population trends. Similarly, hurricanes hardly impact population dynamics in growing, low-density counties.

We find evidence that is inconsistent with our hypothesis in high-density, growing counties. In these counties, hurricane events and related losses in the current year and a greater number of hurricanes in the past decade suppress future population growth. However, when hurricanes in the past decade have been very costly, future population growth tends to be greater. In other words, while hurricane events and losses tend to suppress population growth in counties with growing, high-density populations, this effect may be reversed when substantial hurricane-related monetary losses over the past decade have promoted investment in those places, consistent with the "recovery machine" thesis (Pais and Elliott 2008). These results begin to reconcile the inconsistencies in the sparse literature on hurricane and hazard impacts on population, but further research is needed to understand where and when these effects promote or suppress population growth. For now, we conclude that hurricanes are heterogeneous in their population impacts, and that past population trends and cumulative hurricane events and losses contribute to this heterogeneity.

Our results provide an important corrective to previous investigations into the effects of hurricanes on future growth, which have selected only places with extreme hurricane impacts or have failed to control for the effects of long-term population and hazard event trends. In general, the impact of a single hurricane should be assessed relative to past trends to discern its net effect on future growth. Further, the impact of a single hurricane should be considered together with the cumulative impacts of past hurricanes and relative to counties that have not experienced hurricanes. This quasi-experimental approach leads to generalizable conclusions about the impacts of hurricanes and tropical storms and can be applied to the study of other types of hazards.

Our analysis has several limitations. First, measures of county-level demographic, economic, and environmental characteristics would have been preferable to a simple measure of past population trends and population density, but they were not consistently available in all county-years. During this period, county boundaries changed, and counties' demographic, economic, and political characteristics have changed in ways that were difficult to measure consistently across time and space. To make the project tractable and demonstrate the value of proceeding in this line of inquiry, we used simple metrics and a model that takes into account this unobserved heterogeneity. Second, our multivariate analysis did not take account of the spatial relationships between counties. While a spatial regression analysis is planned, in this analysis we sought to explore new measures and their use in a multivariate regression framework. Nevertheless, our systematic approach to spatial-temporal data shows that hurricanes have heterogeneous impacts on counties, and that their impact depends on past population change and population density, thereby explaining some of the inconsistencies in the growing field of research on weather-related hazards and population change.

Notes

1. The spatial boundary files used for our mapping and modeling were generated from the 2010 TIGER/Line Shapefile. Historical SHELDUS data have already been conflated to modern (2010) boundaries and thus require no boundary corrections over time. The population data used in our study, the U.S. Census Bureau county-level intercensal population estimates, are based on decadal boundaries that are anchored on boundary definitions at the end of the decade. In other words, we had to correct for boundary issues only in 1970, 1980, 1990, and 2000, not all individual years. Most county boundaries do not change. For those that have changed, we use the basic, but standard, process of areal weighting, or reassigning population counts based on the proportion of the county area that changed.

2. h_i induces the variation of the parameters across individual counties; α_{it} is the constant; u_i is a group-specific random element, similar to ε_{it} except that for each group there is just a single draw that enters the regression identically in each period.

3. There is considerable debate about the appropriate application of fixed and random effects models for estimating panel data results. A recent paper by Clark and Linzer (2015) sheds light on the debate, offers guidance on choice of models, and encourages practical and theoretical assessments. We do not estimate a fixed-effects model because we have theoretical and practical reasons to include county-level effects, and a fixed-effects model would make evaluating these effects impossible (Clark and Linzer 2015, 407). In this article, we present the results of our random effects estimation that corrects the standard errors. Furthermore, we evaluated models with a robust correction to control for heteroskedasticity.

4. All losses are standardized to $2014 values.

5. HVRI. 1960–2014 U.S. Hazard Losses; 2008 U.S. Hazard Losses. Available from http://hvri.geog.sc.edu/SHELDUS/.

6. A better measure would be losses relative to the value of the property and crops at risk, but such a measure is not easily obtained. See Ash, Cutter, and Emrich (2013) for a measure of the relative loss ratio using SHELDUS and other county-level data.

References

Ash, Kevin D., Susan L. Cutter, and Christopher T. Emrich. 2013. Acceptable losses? The relative impacts of natural hazards in the United States, 1980–2009. *International Journal of Disaster Risk Reduction* 5:61–72.

Bilsborrow, Richard. E. 1992. *Rural poverty, migration, and the environment in developing countries: Three case studies*, vol. 1017. Washington, DC: World Bank Publications.

Black, Richard, W. Neil Adger, Nigel W. Arnell, Stefan Dercon, Andrew Geddes, and David S. G. Thomas. 2011. The effect of environmental change on human migration. *Global Environmental Change* 21: S3–S11.

Clark, Tom S., and Drew A. Linzer. 2015. Should I use fixed or random effects? *Political Science Research and Methods* 3 (2): 399–408.

Cross, John A. 2014. Disaster consequences of U.S. communities: Long-term demographic consequences. *Environmental Hazards* 13 (1): 73–91.

Davis, Kingsley, and Mikhail Bernstam, eds. 1991. *Resources, environment, and population*. New York, NY: Oxford University Press.

de Sherbinin, Alex, David Carr, Susan Cassels, and Leiwen Jiang. 2007. Population and environment. *Annual Review of Environment and Resources* 32:345–73.

de Sherbinin, Alex, Leah K. VanWey, Kendra McSweeney, Rimjhim Aggarwal, Alisson Barbieri, Sabine Henry, Lori M. Hunter, Wayne Twine, and Robert Walker. 2008. Rural household demographics, livelihoods and the environment. *Global Environmental Change* 18:38–53.

Elliott, James R. 2014. Natural hazards and residential mobility: General patterns and racially unequal outcomes in the United States. *Social Forces* 93 (4): 1723–47.

Gall, Melanie, Khai Hoan Nguyen, and Susan L. Cutter. 2015. Integrated research on disaster risk: Is it really integrated? *International Journal of Disaster Risk Reduction* 12:255–67.

Gutmann, Myron P., and Vincenzo Field. 2010. Katrina in Context: Environment and migration in the U.S. *Population and Environment* 31:3–19.

HVRI. 2015. Spatial hazard events and losses database for the United States, Version 14.1. [Online database]. Columbia, SC: Hazards and Vulnerability Research Institute, University of South Carolina.

Hugo, Graeme, ed. 2013. *Migration and climate change*. New York, NY: Edward Elgar Publishing.

Hunter, Lori M., Jessie K. Luna, and Rachel M. Norton. 2015. Environmental dimensions of migration. *Annual Review of Sociology* 41:377–97.

Intergovernmental Panel on Climate Change (IPCC). 2014. Summary for policymakers. In *Climate change 2014: Impacts, adaptation, and vulnerability. Part A: Global and sectoral aspects*. Contribution of Working Group II to the Fifth Assessment Report of the Intergovernmental Panel on Climate Change, eds. C. B. Field, V. R. Barros, D. J. Dokken, K. J. Mach, M. D. Mastrandrea, T. E. Bilir, M. Chatterjee, K. L. Ebi, Y. O. Estrada, R. C. Genova, B. Girma, E. S. Kissel, A. N. Levy, S. MacCracken, P. R. Mastrandrea, and L. L. White, 1–32. New York, NY: Cambridge University Press. Available from http://ipcc-wg2.gov/AR5/images/uploads/WG2AR5_SPM_FINAL.pdf.

Logan, John R., Sukriti Issar, and Zengwang Xu. 2016. Trapped in place? Segmented resilience to hurricanes in the Gulf South. *Demography* 53:1511–34.

Morss, Rebecca E., Olga V. Wilhelmi, Gerald A. Meehl, and Lisa Dilling. 2011. Improving societal outcomes of extreme weather in a changing climate: An integrated perspective. *Annual Review of Environmental Resources* 36:1–25.

Myers, Norman. 2002. Environmental refugees: A growing phenomenon of the 21st century. *Philosophical Transactions: Biological Sciences* 357 (1420): 609–13.

NOAA. 2013. *National coastal population report: Population trends from 1970 to 2020*. Washington, DC: NOAA. Available from: http://oceanservice.noaa.gov/facts/coastal-population-report.pdf.

Pais, Jeremy, and James R. Elliott. 2008. Places as recovery machines: Vulnerability and neighborhood change after major hurricanes. *Social Forces* 86:1415–53.

Schultz, Jessica, and James R. Elliott. 2013. Natural disasters and local demographic change in the United States. *Population and Environment* 34:293–312.

Shumway, J. Matthew, Samuel Otterstrom, and Sonya Glavac. 2014. Environmental hazards as disamenities: Selective migration and income change in the United States from 2000–2010. *Annals of the Association of American Geographers* 104 (2): 280–91.

U.S. Census Bureau. 2016. Population Estimates, Pastal Data. Washington, DC: U.S. Census Bureau. Available from http://www.census.gov/popest/data/pastal/index.html.

Wright, James D., Peter H. Rossi, Sonia R. Wright, and Eleanor Weber-Burdin. 1979. *After the clean-up: Long-range effects of natural disasters*. Beverly Hills, CA: Sage Publications.

New Trends and Patterns in Western European Immigration to the United States: Linking European and American Databases

By
ELYAKIM KISLEV

This study explores the latest changes in Western European immigration to the United States by integrating several large databases: the U.S. census, the American Community Surveys, the European Social Survey, as well as the Human Development Index and Gini index. Findings show that the number of individuals born in Western Europe but with family origins elsewhere who have been immigrating to and settling in the United States is increasing. I divide the Western European population that immigrates to the United States into seven different subpopulations by their ancestries and explore the characteristics of these populations before and after immigrating to the United States. I also examine their relative success in terms of economic and labor outcomes in America, finding, for example, that some of the least advantaged immigrant groups have some of the best economic outcomes in the United States. The different self-selection and assimilation patterns among these immigrants have implications for U.S. public policy, which we identify and begin to explore.

Keywords: immigration; minorities; integration; Western Europe

In the second decade of the twenty-first century, net migration inflows to Western Europe per year are almost ten times higher than they were in the 1960s (Castles, de Haas, and Miller 2013). These inflows to Western Europe are also changing the demographic composition of Western European immigration to the United States, as first- and second-generation immigrants are starting to move to other destinations. Because the United States is a top receiving country for Western European immigrants, these changes

Elyakim Kislev is a professor at the Federmann School of Public Policy and Government at the Hebrew University. He is the coauthor and coeditor of two books on minorities and social justice and his articles focus on immigration, minorities, family issues, and social policy.

Correspondence: elyakim.kislev@mail.huji.ac.il

DOI: 10.1177/0002716216682692

will affect American society. Furthermore, given the rise in anti-immigrant sentiment in Europe, and following the recent flows of refugees to Western Europe, the number of non-European immigrants arriving in the United States from Western Europe as a second immigration is expected to rise even more. These demographic changes have important implications for the labor market due to the different skill sets of migrants and due to the import of cultures that adhere to traditional gender roles.

The following sections address various facets of the changing nature of the Western European migration to the United States by integrating several data sources. By integrating various databases to examine the different origin groups within Western Europe, we can compare and explore their characteristics and trajectories after migration. This design helps in identifying self-selection patterns and explicating the effects of Western European origins. Thus, the main question of this study is, Do Western European U.S. immigrants' characteristics and trajectories differ by whether they are native Western Europeans or were previously immigrants to Western Europe from other countries?

The Western European immigrant populations that immigrated to the United States include sub-Saharan Africans, North Africans and Middle Easterners (MENA), South Asians, Southern Europeans, Eastern Europeans, internal Western European immigrants (individuals whose parents were born in another Western European country), and the native Western European population. Additionally, these groups are compared to the native-born American population. Western Europe is defined here as: Austria, Belgium, Denmark, Finland, France, Germany, Ireland, Luxemburg, the Netherlands, Norway, Sweden, Switzerland, and the United Kingdom (UK).

Immigrants in Western Europe

Being a minority in the European labor market is associated with a relative disadvantage or "ethnic penalty" (Heath and Ridge 1983). Ethnic penalties among minorities have been reported in many studies, reflected by higher rates of unemployment, lower salaries, and reduced opportunity for promotion at work (e.g. Röder and Mühlau 2011; Safi 2010; Becker 1971; Constant, Kahanec, and Zimmermann 2009).

Mounting evidence suggests that the nature and severity of the relative disadvantages vary across ethnic groups and national context. For example, one study depicts a clear hierarchy across Western Europe in rates of unemployment among minorities (Heath and Cheung 2007). The results indicate that Southern Europeans in Western Europe exhibit levels of unemployment that are similar or equal to that of the native population. Higher levels of unemployment than Southern Europeans are exhibited by Eastern Europeans, Chinese, and Indian minorities, and finally all other non-European minorities, in this order. In some contexts, minorities from the Middle East and North Africa face a particularly heavy ethnic penalty and perform even less well. Other studies have found evidence of similar social hierarchies at the national level. In the UK, for instance, Heath and Cheung (2006) demonstrate that minorities of Bangladeshi, Pakistani, and sub-Saharan African descent suffer from the heaviest ethnic penalties in the labor market. In Sweden, Jonson (2007) found the rate of unemployment to be strongly correlated with the level of visibility of the minority group. Similar studies have also demonstrated hierarchical ethnic penalties in other national contexts, for example in France (Aeberhardt et al. 2010), and in the Netherlands (Crul and Vermeulen 2003).

Somewhat similar results are found in the educational system. Heath, Rothon, and Kilpi (2008) state that minorities whose ancestries are from less-developed countries tend to exhibit lower educational attainment and qualifications than the native population. Turks, North Africans, Pakistanis, and Caribbean minorities are the most vulnerable in this sense. In contrast, Southern Europeans, Yugoslavians, Chinese, and Indians tend to integrate quite fast and in some cases even overtake natives in terms of educational achievement.

As for gender gaps, studies show that women of all backgrounds enjoy equality in the Western European educational system, and minorities even tend to fare slightly better in higher education (Heath, Rothon, and Kilpi 2008). However, this is not true in the labor market. When looking at a cross-national analysis (Heath 2007), we can see that some groups retain gender gaps in the labor market even among the second generation (though in lower levels than the first-generation). Women of Moroccan, Pakistani, or Turkish ancestry tend to have lower participation rates in the labor market than the parallel majority group in the same country. Research in Denmark (Nielsen et al. 2003) shows that although educational achievements are equal, second-generation women find it more difficult to find a job, maintain the job, and receive an equal wage.

Attitudes regarding immigration may also affect the ethnic penalties that minorities face in the labor market. Indeed, the past two decades have seen the emergence of a robust literature addressing social attitudes toward immigrants and immigration (Ceobanu and Escandell 2010; Polyakova 2015; Wodak and Boukala 2015). In seeking to explain the social mechanisms behind discrimination against minorities, the literature points to a steady rise in anti-immigrant sentiment across most, if not all, of Western Europe. This trend is exemplified by Semyonov, Raijman, and Gorodzeisky's (2006) study that compared patterns in antiforeign sentiment between 1988 and 2000. Using data from the Eurobarometer, the study investigated trends in attitudes toward immigrant population size as well as the perceived impact that foreigners have on the welfare system, unemployment, and delinquency/violence. The results pointed to a clear rise in xenophobia across Western Europe.

There are cultural and demographic dissimilarities between the Western European majority group and Western European minorities. These different characteristics might also affect the character of emigration flows from Western Europe to the United States. Indeed, minorities tend to preserve their own way of living, which is often measured by language acquisition, birthrate, mean age of marriage, and mean age of first childbearing (Heath, Rothon, and Kilpi 2008).

A study conducted in the Netherlands (Garssen and Nicolaas 2008) shows that Moroccan and Turkish immigrants have a much higher birthrate than the native population, even higher than in the country of origin. For example, in 1990, Moroccans in the Netherlands had 5 children per family, Turks had 3, and the native population had 1.7. Although this gap declines over time, and second-generation immigrants have lower fertility rates than first-generation, there is still a gap. The authors also show that the mean age of first childbearing among Moroccans is 26.5, among Turks it is 24.5, whereas among the native population it is 29.5. The same holds true in the UK (Coleman, Compton, and Salt 2002), where the total fertility rate in 1996 was 1.7 among the native population, 4.9 among Pakistanis and Bangladeshis, 2.25 among Indians, and among Africans it was 4 children per family.

Western European minorities' immigration to the United States

The main question of this study is whether the cumulative disadvantage of Western European minorities affects their patterns in settling and integrating into American society. As shown above, there is a cumulative disadvantage among minorities within Western Europe. They tend to have lower educational achievements. Among those who do graduate, there are still higher chances of unemployment. Even among those who find a job, there is inequality in admission into prestigious jobs and access to an equal salary. Western European minorities who immigrate to the United States might be affected by these difficulties in Western Europe. Cohen (1996) and Kislev (2014) show that Israeli minorities carry lower levels of human capital upon arrival to the United States and suggest that it is due to the social exclusion they experience.

Moreover, Kislev (2014) shows that there are dissimilarities among different Israeli minorities; this might also be the case for Western European minority groups. Groups that experience higher social exclusion might carry a higher ethnic penalty. Since social exclusion is determined based on distance in terms of social, cultural, and economic factors (Fossett 2006, Parrillo and Donoghue 2005), Western European minorities originating in less-developed countries may have lower achievements upon arrival. In contrast, Western European minorities whose parents came from within Europe may exhibit higher achievements.

In this sense, selection mechanisms might also intervene and affect the gap in human capital between minorities and the majority group (Borjas 1987). A positive selection is expected because immigration is more accessible among the highly skilled and wealthier minority members (Kislev 2014). Thus, those with high levels of human capital in Western Europe are expected to be the ones who actually immigrate to the United States, while the lower strata of minorities in Western Europe do not immigrate because they cannot afford to do so. Selection mechanisms might entail higher levels of human capital among minorities who immigrate to the United States than the average human capital levels among those living in Western Europe. Moreover, some highly skilled minorities might have more incentive to immigrate since they experience not only wage discrimination but also hiring discrimination; European majority members tend to hire members of their own group for more prestigious jobs and tend to pay them better (Aigner and Cain 1977; Kogan 2006; Van Tubergen, Maas, and Flap 2004).

Thus, self-selection processes may be important mechanisms in determining Western European minorities' human capital upon arrival to the United States and how it might affect their integration thereafter. These mechanisms are examined below. We also hypothesize that Western European minorities show similar demographic characteristics to those that they had in the country of origin, which are different from those held by the majority.

Finally, the pace of economic integration might also be different for each group, though the nature and extent of this difference is an empirical question that should be tested. Indeed, the common benchmark to measure immigrant economic advancement is the question of how different immigrants are from American-born citizens in terms of socioeconomic standing, language use, and intermarriage (Borjas 2001). But immigrant minorities need to be compared mainly to the parallel majority group from the same country. This comparison shows how their characteristics and integration patterns differ from those who came from what appears to be the same economic and political system. Thus, the main question addressed here is whether integration of European migrants into the United States is different for native Europeans compared with those who were previous immigrants to Europe from other countries.

Data and Methods

To examine this question, this study integrates several large databases. First, this research uses data from the European Social Surveys (ESS) to investigate the groups within Western Europe before migration to the United States. The ESS has data from 2002 to 2012 in intervals of two years. We used 2004–2012 ESS data to examine second-generation Western European minorities because the ESS provides specific parents' country of birth from only 2004 onward. The population and design weights provided by the ESS were incorporated into all calculations.

Two algorithms were applied to identify the populations under investigation in the ESS. The primary one identified respondents whose parents were born in a foreign country (e.g., Tunisia). These respondents are identified as second-generation immigrants to Western Europe (e.g., of North African origins). In addition, because some studies define second-generation immigrants as those with both parents born in the country of origin, the analyses were also conducted according to this latter definition. The results do not differ. Note that North Africans and Middle Easterners were combined because initial analyses showed a high similarity between these two groups.

The findings on Western European migration to the United States are mainly based on data from the 2000 U.S. Census and 2001–2013 American Community Surveys (ACS; Ruggles et al. 2013). Additional databases have been part of this research, such as the New Immigrant Survey, which is conducted among those applying for permanent residency in the United States; and the International Passenger Survey, which is conducted among UK international passengers. However, such attempts did not succeed in obtaining the necessary information or in identifying the populations under investigation with sufficient accuracy and enough observations.

In identifying the populations in the American databases, two algorithms were applied together. First, Western European–born immigrants were chosen on the basis of the birthplace variable given in the census and surveys (Western European countries). Though many divisions and subdivisions could be used to define the borders of Western Europe, Western Europe is defined in this study as including Austria, Belgium, Denmark, Finland, France, Germany, Ireland, Luxemburg, the Netherlands, Norway, Sweden, Switzerland, and the United Kingdom. Note that despite the similarities between these countries of Western Europe, the cross-classified analyses presented below also control for the specific Western European country of origin as another level.

Based on the works of Cohen and Tyree (1994), Cohen and Haberfeld (1997), and Kislev (2014, 2015), we use the ancestry variable as our second step to identify the different Western European immigrant groups. For instance, individuals of South Asian ancestry (e.g., Indian) who were born in Western Europe are identified as Western European South Asian immigrants to the United States. Note that the answer to the ancestry question has been employed extensively in various studies to classify ethnic identification and ethnicity (Lieberson and Waters 1985; Lieberson 1985; Neidert and Farley 1985; Kislev 2015). To maintain accuracy in conclusions, this study excludes all respondents who report being "born abroad to American parents."

Since the focus of this study lies in the economic integration of working-age adults, the samples are restricted to men and women between the ages of 20 and 54. To account for the "1.5 generation" (Zhou 1997), a dummy variable in the multivariate analysis is included for individuals who immigrated to the United States at age 15 or younger. Note that this variable is not consistently significant. All calculations are also weighted appropriately.

To account for differences between the different origin Western European countries, measures of the Gini index by the World Bank (2013) database are incorporated and in the rare cases where data were not available for a specific country, measures were taken from the CIA database (2013). In addition, measures of human development were taken from the United Nation's (UN) Human Development Index (HDI) database (UN 2013). The HDI combines measurements of a country's life expectancies; adult literacy rates; GDP; and gross enrollment ratios in primary, secondary, and tertiary education.

To find self-selection mechanisms in migration, the results from American sources are compared with the parallel European sources. This study focuses on four main variables that reflect groups' self-selection and are comparable between the different databases: age, unemployment rates, percentage of participation in the labor force, and percentage of BA holders. In addition, this research presents findings of the social stance of these groups within Western Europe and uses the latest International Social Survey Programme (ISSP Research Group 2013) to compare across major countries in Western Europe and the United States.

To understand the characteristics of immigrant minorities from Western Europe in the United States compared to the Western European majority group and the native-born American population, this study uses variables that capture the most relevant traits such as age, education, and tenure in the country. The income variables were adjusted to the inflation rate of each census for the preceding year.

The key independent variable for the test of economic advancement over the years is "years spent in the U.S." in interaction with a specific ethnic group within the Western European immigrant group. To measure economic integration, this study focuses on three dependent variables: household income, total personal income, and wage income. These economic outcomes were logged and standardized to provide tangible comparisons between the different measures and a deeper understanding of the results. Naturally, this does not affect the results in any way.

This study accounts for variation due to both origin country in Western Europe and destination state in the United States by conducting a cross-classified multilevel analysis. The origin countries and destination states are incorporated in levels 2 and 3. Hence, the research presented here can be used to analyze Western European immigration to the United States nested separately within both destination states in the United States and origin countries in Western Europe (see Van Tubergen, Maas, and Flap 2004).

The cross-classified multilevel analyses also incorporate measures of the educational and economic levels of the ethnic group in the Western European country of birth (Gans 2007; Alba and Nee 1997, 2005; Cohen 1996; Kislev 2014; Borjas 1982). To ensure accuracy, data on the origin ethnic group in Western Europe are drawn only from groups that have more than thirty respondents in the ESS. The average educational level is measured as the percentage of BA holders among the ethnic group in Western Europe relative to the overall percentage of BA holders in each specific Western European country. The average income level is measured as the average level (on a scale) of household income among the ethnic group in Western Europe relative to the average income level of each specific Western European country. Finally, the Gini index and HDI are incorporated.

Results

The characteristics of the origin groups

To understand the differentiation among Western European immigrants to the United States, we need to survey their characteristics within Western Europe. Figure 1 presents the percentages of feelings of discrimination among each ethnic group and includes, for a comparative look, both first- and second-generation immigrants in Western Europe. This figure shows a clear stratification in feelings of discrimination, where Western European minorities who have non-European origins feel the most discriminated against.

To compare these results with the social condition of the same ethnicities in the United States, findings from the ISSP (ISSP Research Group 2013) indicate that the general condition of these minorities might be better in the United States. Forty-six percent of Americans think that immigrants are generally good for the country's economy. In contrast, only 22 percent, 24 percent, and 30 percent of respondents agree with this statement in the UK, France, and Germany, respectively. Similarly, although 58 percent of Americans think that immigrants improve society, only 33 percent, 35 percent, and 49 percent believe the same in

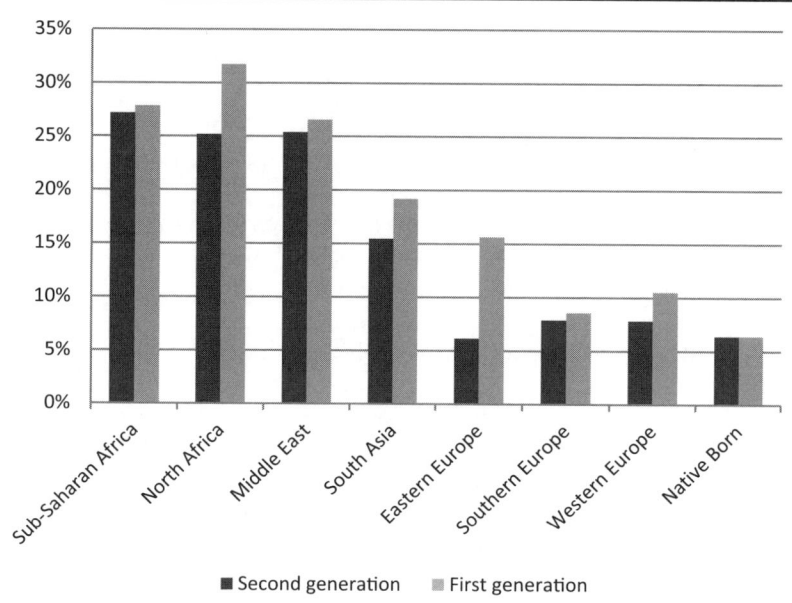

FIGURE 1
Percentage of Those Feeling Discriminated against within Western Europe, by Ethnicity, Age 20–54

SOURCE: European Social Surveys (2004–2012).
NOTE: Calculations were made with weights.

the UK, France, and Germany respectively. A similar pattern is observed with perceived crime rates, with 25 percent of Americans agreeing that immigrants increase crime rates, and 38, 42, and 60 percent of respondents saying the same in the UK, France, and Germany, respectively.

The findings presented in Figure 1 correlate with economic measures among these groups as shown in Table 1. The columns of the findings from within Western Europe show that the most discriminated against groups are those who are disadvantaged economically as well.

In addition, Table 1 sheds light on some self-selection patterns. In terms of educational self-selection, three groups show strong selection processes: men of North African and Middle Eastern origins as well as both men and women of Southern European origins. In general, those who immigrate to the United States are highly educated relative to their parallel groups in Western Europe.

The most striking finding in Table 1, however, is the low unemployment rates among North Africans, Middle Easterners, and South Asians in the United States. All three groups show high unemployment rates in Western Europe in comparison to the majority group. However, after coming to the United States, in the first 10 years of their tenure, they find the labor market much more accessible. Sub-Saharan Africans, however, continue to show high rates of unemployment, especially sub-Saharan African men.

TABLE 1
Western European Immigrants in the United States (US) for Fewer than 10 Years versus the Origin Groups (2nd Generation Immigrants) in Western Europe (WE), Age 20 to 54

Place of origin	Gender	Average age In WE	Average age In the US	% in labor force In WE	% in labor force In the US	Unemployed[a] In WE	Unemployed[a] In the US	Have a BA In WE	Have a BA In the US
Sub-Saharan Africa	Female	33.32	32.67	81.71%	70.38%	13.43%	6.93%	29.27%	52.61%
	Male	32.89	31.14	87.83%	74.91%	8.91%	10.19%	24.56%	47.64%
MENA[b]	Female	33.14	30.35	76.74%	54.30%	20.45%	9.17%	26.16%	53.39%
	Male	32.66	30.43	93.30%	79.51%	17.68%	5.15%	18.13%	61.07%
South Asia	Female	34.64	30.77	81.36%	57.62%	6.25%	4.96%	44.07%	65.24%
	Male	33.89	30.54	94.00%	84.24%	17.02%	3.74%	36.00%	67.62%
Eastern Europe	Female	39.78	35.31	81.82%	65.67%	6.11%	6.54%	25.45%	45.92%
	Male	38.30	36.08	84.54%	84.51%	5.28%	2.09%	25.09%	56.64%
Southern Europe	Female	37.92	34.25	76.88%	63.16%	8.39%	5.30%	17.20%	49.76%
	Male	37.87	33.82	91.71%	86.05%	10.11%	4.95%	19.51%	53.88%
Western Europe (nonnatives)	Female	40.98	35.80	81.08%	61.93%	8.33%	3.96%	33.78%	51.33%
	Male	40.23	36.45	90.28%	90.16%	3.85%	3.93%	27.78%	61.96%
Native Western European	Female	38.38	34.85	78.06%	58.52%	7.84%	5.36%	33.49%	49.84%
	Male	38.54	35.78	89.87%	87.95%	7.30%	2.82%	31.23%	65.77%

SOURCE: Data from 2000 U.S. Census, 2001–2013 ACS, and 2002–2012 ESS.
a. Of those in labor force.
b. Middle East and North Africa.

Finally, Table 1 also shows that Western European immigrants who come to the United States are younger on average compared with their counterparts who stay in Western Europe. There is no particular difference between the groups in this sense. In addition, one can see lower participation rates in the labor markets across all groups, but this is due to the high number who are in school.

The latter comparison leads us to ask, How many of these groups come to the United States to acquire higher education? This question is particularly important since, as described earlier, the educational achievements of minorities within Western Europe are lower and discrimination is prevalent. Indeed, the data suggest that non-European immigrants from Western Europe are both more educated and more likely to be enrolled in school in comparison with other groups. These findings hold in a cross-classified multilevel analysis that controls for demographic

TABLE 2
Cross-Classified Multilevel Analyses of Education Characteristics for Immigrants to the United States, Age 20 to 54

	Men		Women	
Variable	Have BA	In school	Have BA	In school
Age	0.998°	0.857°°°	0.984°°°	0.895°°°
Place of origin[a]	0.539°°°	1.661°°°	1.165°°	1.742°°°
Sub-Saharan Africa				
MENA[b]	1.122	1.773°°°	1.349°°°	1.475°°°
South Asia	1.701°°°	1.486°°°	2.161°°°	1.330°°°
Eastern Europe	0.731°°°	1.421°°°	1.295°°°	1.417°°°
Southern Europe	0.474°°°	0.96	0.783°°°	0.973
Western Europe	0.835°°°	1.061	1.152°°°	1.125°°
Intercept	1.279°°°	22.549°°°	1.152	6.501°°°
SD components	0.276°°°	0.241°°°	0.332°°°	0.174°°°
Birth place SD				
State SD	0.400°°°	0.199°°°	0.367°°°	0.125°°°
N	58,552	58,552	62,960	62,960

°$p < .1$ °°$p < .05$. °°°$p < .01$ (two-tailed test).
SOURCE: Data from 2000 U.S. Census and 2001–2013 ACS.
a. Reference category: Native Western European.
b. Middle East and North Africa.

and geographic variables such as age, gender, birthplace in Western Europe, and region within the United States to which they immigrate (see Table 2).

Table 2 also shows that women who are not native Western Europeans are generally more educated than Western European natives who immigrate to the United States. This excludes Southern European women, who are less educated than native Western European women. Among men, however, the situation is different. Native Western European immigrants are more educated than most groups except for Middle Eastern, North African, and South Asian men.

Ethnic discrepancies among Western European immigrants to the United States

Table 3 presents the demographic characteristics of the seven immigrant groups under investigation in comparison with native-born Americans. The analysis of these characteristics shows that minority immigrants whose parents are from non-European countries are generally younger than other Western European immigrant groups and native-born Americans. They also show shorter tenure in the United States. Both findings are not surprising considering their late arrival to the United States. It is interesting to note that many Eastern Europeans arrived as children. This is consistent with a refugee profile—many Eastern Europeans escaped Western Europe during and immediately after

TABLE 3
Demographic Characteristics of Western European Immigrants, Age 20 to 54

Place of origin	Gender	N	Mean age	Mean age at arrival	% arrived as a child (≤15)	Mean years in the USA	% married	Mean number of children[b]	% naturalized citizens
Sub-Saharan Africa	Female	1037	36.68	16.47	44.07%	20.20	45.42%	1.11	52.75%
	Male	935	35.71	16.71	45.35%	19.00	43.64%	0.71	51.02%
MENA[a]	Female	549	33.73	18.41	32.97%	15.32	56.83%	0.90	47.91%
	Male	603	33.61	18.84	32.67%	14.77	47.93%	0.51	46.27%
South Asia	Female	1299	32.70	16.15	44.80%	16.54	64.67%	0.91	53.35%
	Male	1200	32.89	15.08	49.83%	17.81	54.33%	0.67	57.25%
Eastern Europe	Female	1924	45.71	10.53	71.15%	35.18	66.48%	0.83	76.56%
	Male	1865	46.22	9.93	73.24%	36.29	68.15%	0.90	75.82%
Southern Europe	Female	922	38.71	15.15	52.06%	23.56	65.51%	1.12	56.29%
	Male	903	38.58	16.19	47.29%	22.39	68.00%	0.97	51.61%
Western Europe (non-natives)	Female	4,914	41.32	18.47	38.63%	22.85	70.74%	1.06	43.14%
	Male	4,321	40.67	19.60	36.44%	21.07	68.53%	0.84	39.49%
Native Western European	Female	53,316	40.22	21.75	24.49%	18.47	70.19%	0.98	35.29%
	Male	40,330	39.71	23.07	24.28%	16.64	68.43%	0.84	31.83%
Native-born Americans	Female	66,154	38.15				56.87%	1.01	
	Male	63,195	37.99				54.92%	0.82	4

SOURCE: Data from 2000 U.S. Census and 2001–2013 ACS.
a. Middle East and North Africa.
b. Of those who are/were married.

World War II and remained in the United States, which has also led to higher naturalization rates overall. In contrast, the Western European majority group tends to acquire/receive citizenship in lower numbers, especially among women.

The low average age of non–Western Europeans somewhat explains their lower marriage rate. However, South Asians show a high marriage rate despite being among the youngest group of Western European immigrants. There are also differences in fertility rates. Sub-Saharan Africans and Southern Europeans are at the top in this sense.

Table 4 again demonstrates the importance of distinguishing between the different immigrant groups by presenting economic indicators. In general, native Western Europeans are in a far better economic position than all other groups. They demonstrate higher household income, total income, wage income, employment rates, labor force participation; and more prestigious jobs as reflected in the socioeconomic index. In contrast, non-European immigrants who were born in Western Europe stand at the other end. They show lower income levels as well as lower labor market participation rates.

Table 5 examines yet another facet of the differences between the groups by comparing their educational and geographic characteristics. Non-European minorities are distinguished in education again by the fact that a higher percentage of them are in school. A higher number hold a BA degree, although they are less fluent in English. These seven groups concentrate in different geographic regions within the United States. Sub-Saharan Africans have a strong tendency to immigrate to the South. In contrast, Middle Easterners and North Africans concentrate in the West in higher numbers. South Asians are similar in this sense, but also immigrate to the South. Eastern Europeans do not concentrate in these two destinations, but rather immigrate heavily to the Northeast and the Midwest. Southern Europeans are highly concentrated in the Northeast. Finally, relative to native-born Americans, Western Europeans (both natives and non-natives) tend to immigrate to the West and to the Northeast and less to the South and the Midwest. Figure 2 demonstrates this division.

The growing migration of minorities from Western Europe to the United States

Figure 3 illustrates the proportion of each incoming ethnic group in each year relative to its total migration in all years. Figure 3 shows that after World War II, there was a large influx of immigrants from Western Europe whose ancestry was not from the country of birth. This influx probably included many refugees who escaped to the United States. In subsequent years, however, the share of non-native European immigrants decreased. In contrast, there is a general growing trend of non-European immigrants who come to the United States from Western Europe.

The data also show that second generation sub-Saharan Africans, Middle Easterners, North Africans, and South Asians coming from Western Europe to the United States and naturalizing has increased to around 50 percent in the 2000s from only 10–15 percent in the 1950s. However, immigration of Eastern and Southern Europeans coming from Western Europe has hardly increased. These numbers together with Figure 3 clearly show that the number of Western European minorities who come to the United States and pursue citizenship and settle down here is continuously increasing.

Economic advancement of Western European immigrants

Table 6 focuses on four main indicators and tests them in cross-classified multilevel analyses. The first indicator is how the initial economic achievements of a

TABLE 4
Economic Characteristics of Western European Immigrants, Age 20 to 54

Place of origin	Gender	% in labor force	Unemployed	Total income Median	Total income Mean	Wage income[b] Median	Wage income[b] Mean	Household income Median	Household income Mean	Socioeconomic index[a]
Sub-Saharan Africa	Female	78.30%	6.16%	24000	31496	28042	33751	43440	56241	38.96	41.83
	Male	79.79%	8.85%	26443	38189	30000	40891	47921	58276	38.51	40.68
MENA[c]	Female	64.48%	7.34%	12038	26593	27552	36837	48328	67838	44.13	43.34
	Male	85.07%	5.46%	35000	55915	37040	52334	56861	79288	47.3	45.02
South Asia	Female	69.13%	4.90%	19425	33222	31472	40941	68186	89953	46.67	45.85
	Male	86.17%	4.16%	43363	69077	45840	67134	73665	98848	47.88	49.82
Eastern Europe	Female	75.99%	5.20%	24150	33750	29171	36921	65160	86305	40.29	41.79
	Male	89.49%	4.25%	46650	65717	45750	60590	70376	91081	43.04	42.75
Southern Europe	Female	72.02%	5.27%	19143	26127	25928	30661	52960	67459	38.51	40.73
	Male	89.37%	4.83%	42000	58469	40972	54207	61883	79718	38.51	40.60
Western Europe (non-natives)	Female	71.40%	5.50%	20000	30825	27583	35204	61000	82181	40.29	41.63
	Male	92.33%	3.31%	49560	68778	46737	63260	70000	87807	44.4	44.12
Native Western European	Female	68.43%	4.62%	17640	27182	24780	31905	59366	76867	38.51	41.23
	Male	90.35%	3.04%	51212	75522	52038	70908	74441	94702	47.43	45.76
Native-born Americans	Female	76.49%	6.46%	17930	24626	22000	26983	42596	55829	36.09	38.60
	Male	85.62%	7.21%	30000	41760	31734	41133	47905	60622	36.08	37.46

SOURCE: Data from 2000 U.S. census and 2001–2013 ACS.
NOTE: Income is in 2000 U.S. dollars.
a. Of those in labor force.
b. Of those working.
c. Middle East and North Africa.

TABLE 5
Educational and Geographic Characteristics of Western European Immigrants, Age 20 to 54

Place of origin	Gender	% in school	% speaks English "very well"	Linguistically isolated	% BA+	Mean city population	% in the Midwest	% in the Northeast	% in the South	% in the West
Sub-Saharan Africa	Female	19.58%	70.40%	1.83%	47.16%	12229	8.81%	26.93%	48.09%	16.16%
	Male	19.79%	68.13%	2.46%	42.57%	10535	10.42%	25.19%	47.12%	17.26%
MENA[a]	Female	23.86%	20.40%	7.10%	56.28%	10944	13.28%	22.88%	21.22%	42.62%
	Male	24.21%	23.55%	6.30%	62.19%	10116	12.25%	25.00%	24.33%	38.42%
South Asia	Female	20.86%	32.64%	0.92%	64.59%	7331	12.50%	21.04%	32.69%	33.77%
	Male	20.50%	43.00%	1.92%	70.33%	7596	13.18%	20.57%	32.66%	33.59%
Eastern Europe	Female	8.47%	46.93%	3.17%	41.94%	8319	24.27%	36.30%	19.19%	20.24%
	Male	5.95%	56.89%	3.59%	47.45%	7828	24.78%	37.47%	16.04%	21.71%
Southern Europe	Female	10.95%	43.82%	3.47%	41.00%	8895	10.26%	40.94%	25.98%	22.82%
	Male	9.08%	43.30%	4.21%	43.52%	8628	12.33%	43.56%	23.33%	20.78%
Western Europe (non-natives)	Female	9.62%	83.53%	0.67%	41.72%	5051	14.09%	23.83%	33.47%	28.61%
	Male	7.85%	85.70%	0.44%	53.01%	6074	12.54%	24.04%	32.89%	30.53%
Native Western European	Female	9.76%	55.06%	1.52%	41.66%	5535	11.09%	24.52%	33.10%	31.29%
	Male	9.05%	62.81%	3.09%	57.66%	7455	14.22%	24.86%	28.53%	32.38%
Native-born Americans	Female	12.54%	93.05%	0.4%	29.81%	2264	25.09%	17.93%	36.72%	20.26%
	Male	9.73%	92.80%	0.39%	26.91%	2181	25.22%	17.69%	36.13%	20.96%

SOURCE: Data from 2000 U.S. Census and 2001–2013 ACS.
a. Middle East and North Africa.

FIGURE 2
Geographical Distribution of Seven Groups of Western European–Born Immigrants to the United States, by State, All Ages

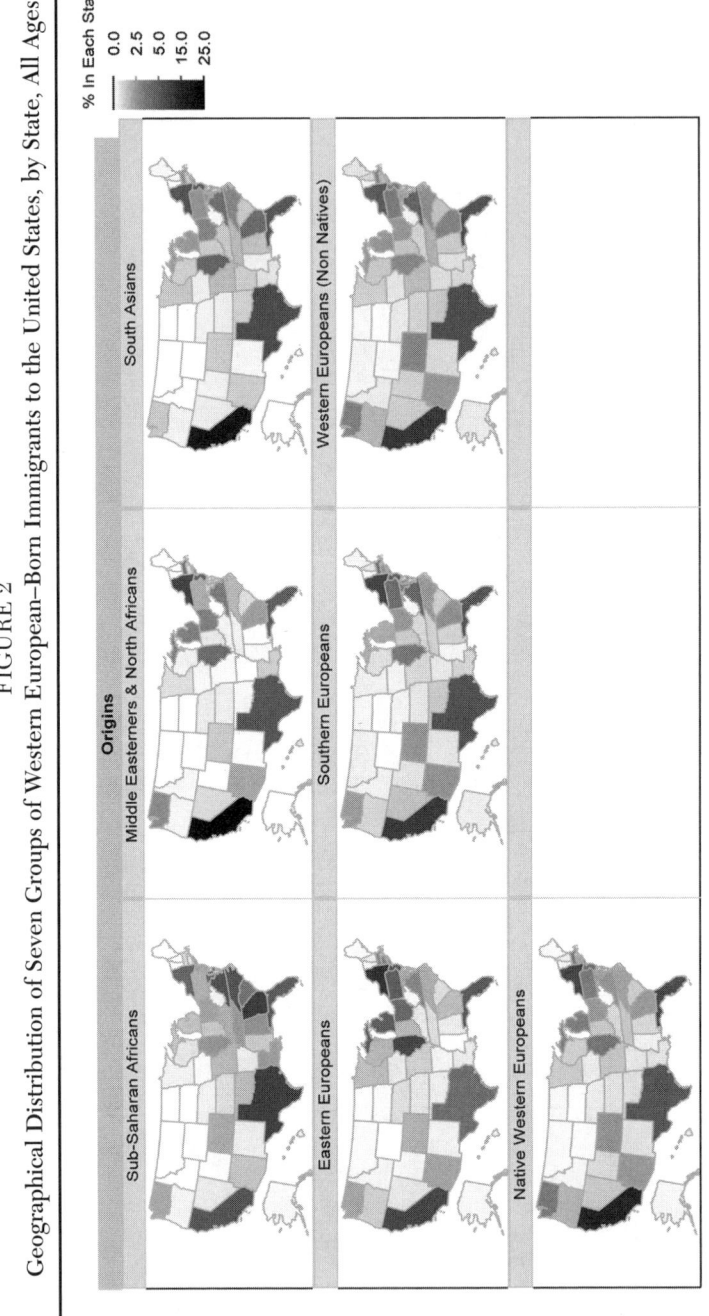

SOURCE: Data from 2000 U.S. Census and 2001–2013 ACS.

FIGURE 3
Trends in Wave Proportions of Migration from Western Europe by Ethnicity, Arrival Age 15 and Above

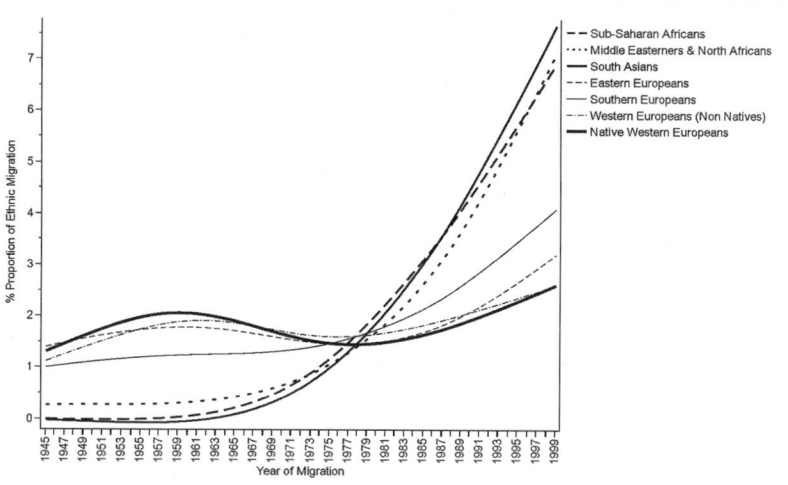

SOURCE: Data from 2000 U.S. Census with 2001–2013 ACS.
NOTE: Calculations were made with weights.

Western European minority group differ from the parallel Western European majority group. In other words, Table 6 shows the economic penalty of each ethnic group (see Freeman 2004; Borgna and Contini 2014; Heath and Cheung 2006). The second important indicator is the pace with which immigrants from Western Europe integrate into the United States for every year that they stay in the country, in comparison to the parallel majority group that migrates from Western Europe. This is done by interacting "years in the U.S." with the specific ethnic group. The third indicator is how the ethnic background of these minorities affects their assimilation in the United States, in terms of the educational and income levels of their origin ethnic groups. The fourth indicator is the effect of the HDI and Gini index of the Western European country of birth.

Table 6 shows that sub-Saharan Africans carry 14 percent of an ethnic penalty in terms of household income while other factors are insignificant. Middle Easterners and North Africans carry 13 percent and 5 percent of an ethnic penalty in terms of total personal income and wage income, respectively. South Asians carry an ethnic penalty of 8 percent in terms of total personal income. Eastern Europeans carry 12 percent of an ethnic penalty in terms of household income. Southern Europeans show an ethnic penalty only in terms of wage income, which stands at 7 percent. In contrast, the condition of Western Europeans whose parents immigrate within Western Europe is much better. They do not show an ethnic penalty relative to those who were born in a country that matches their ancestry.

TABLE 6
Cross-Classified Multilevel Analyses of Economic Characteristics of
Western European Immigrants to the United States, Age 20 to 54

Variable	Household income	Total income	Wage income[d]	Socioeconomic index[e]
Individual characteristics				
Female	0.931***	0.610***	0.734***	0.900***
Age	1.056***	1.145***	1.097***	1.067***
Age squared	0.999***	0.999***	0.999***	0.999***
Not a citizen	0.955***	0.938***	0.974***	0.968***
Hold a BA	1.455***	1.549***	1.335***	2.685***
English proficiency	1.032***	1.042***	1.030***	1.025**
Number of children	1.025***	0.997	0.995	0.986
Married[a]	0.659***	1.006	0.960***	0.949***
Employment	0.713***	0.407***		
Currently in school	0.830***	0.758***	0.810***	1.148***
Arrived as a child (under 15)	0.982	0.99	0.972***	0.986
The effect of ethnic group in W. EU				
Educational level of group	0.996	0.997	0.997	1.009
Income level of group	1.007*	1.008	1.001	1.01
The effect of origin country characteristics				
Gini index of country of birth	0.995	0.997	0.999	1
HDI of country of birth	2.814	4.602***	1.705	2.479**
Place of ethnic origin[b]				
Sub-Saharan Africa	0.857**	1.108	0.988	1.165***
MENA[c]	0.978	0.870**	0.950***	0.934
South Asia	0.98	0.920*	0.965	0.953
Eastern Europe	0.876***	0.988	0.979	0.927*
Southern Europe	0.914	0.979	0.934***	0.861**
Western Europe	0.973	0.998	0.996	1.031
Years in the U.S.	0.996***	0.998***	0.998***	0.996***
Ethnic origins × years in the U.S.[b]				
Sub-Saharan Africa	1.005*	0.990***	0.997	0.998
MENA[c]	1.006***	1.006**	1.003**	1.004
South Asia	1.009***	1.006***	1.004**	1.008***
Eastern Europe	1.004***	1.001	1.001	1.004***
Southern Europe	1.001	0.999	1.001***	1.002
Western Europe	1.001	1	1	0.999
Intercept	0.209**	0.024***	0.073***	0.114***
SD Components				
Birth place SD	0.057***	0.035***	0.021***	0.022***
State SD	0.180***	0.116***	0.078***	0.104***
Residual SD	0.870***	0.819***	0.450***	0.861***
N	83124	75066	60644	65369

*$p < .1$. **$p < .05$. ***$p < .01$ (two-tailed test).
SOURCE: Data from 2000 U.S. Census and 2001–2013 ACS.
NOTE: "Year of survey" variable is included but not shown here.
a. Reference category: Not married (widowed, divorced, single).
b. Reference category: Native Western European.
c. Middle East and North Africa.
d. Of those employed.
e. Of those in the labor force.

As for the economic advancement of these immigrant populations, sub-Saharan Africans advance only in terms of household income, 0.5 percent for each year in the United States, while they actually retreat 1 percent in terms of their personal income. The pace of economic advancement of Middle Easterners and North Africans is 0.6 percent for household and total personal income and 0.3 percent for wage income. The economic advancement of South Asians is 0.9 percent for household income, 0.6 percent for total personal income, and 0.4 percent for wage income. The economic advancement of Eastern Europeans is slower and stands at 0.4 percent for household income while their pace in other economic measures is insignificant. The pace of economic advancement of Southern Europeans is significant only for wage income and stands at a mere 0.1 percent relative to native Western Europeans.

In terms of the effect of the origin ethnic group's characteristics, the only correlation that is positive and significant is the level of household income. The level of household income of individuals who immigrate to the United States is 0.7 percent higher for every one standardized point in the household income level of the parallel ethnic group in Western Europe that matches their ancestry.

Differences between Western European countries are significant only when it comes to the HDI of the Western European country of origin, with two significant coefficients for economic characteristics. Note that origin country and state variations are significant but still do not have a large share in explaining the outcomes. In addition, the variation between states within the United States is generally higher than that of origin countries of Western Europe.

Discussion

Immigrants move, and research surveys do not usually move with them. Because of this, researchers and policy-makers can miss crucial pieces of information such as immigrant self-selection and the effect of origins on economic performance in destination countries. This study addresses the problem of tracking immigrants across countries by integrating the U.S. census; ACS; ESS; and the World Bank, CIA, and UN databases. This study, therefore, provides a wider picture of the characteristics and mechanisms of Western European immigration to the United States.

First, we find that Western European immigration to the United States, especially its settlement patterns, is changing. Western European minorities are acquiring American citizenship in growing numbers, and although a large number of them come to the United States to attend American universities, they tend to stay in high numbers.

Given those findings, we investigated the new composition of Western European immigration to the United States, including the characteristics of newcomers, and how well they integrate into American society and the U.S. labor market. We find that Western European minorities arrive younger and that they are highly selected and educated. Moreover, they hold demographic characteristics that resemble those of their origin ethnic groups in Western Europe, in terms of their marital status and number of children.

Additionally, we find that Western European minorities carry an ethnic penalty in terms of their economic and labor market outcomes in the United States that is not related to the educational and income levels of their origin ethnic group in Western Europe, or to the Gini index and HDI of their Western European country of birth. That economic penalties are directly related to ethnicities (Heath and Ridge 1983; Heath and Cheung 2006; Heath 2007) suggests that human capital theories (Borjas 1987; Chiswick 1978) are not sufficient to explain our findings. More research is required to disentangle the reasons for these ethnic penalties—further research should integrate European and American databases and test more possibilities at both the individual and ethnic levels.

Strikingly, though, we show that immigrants of non-European descent progress faster than their Western European majority counterparts in terms of their economic outcomes in America, despite carrying a heavier ethnic penalty. Again, human capital theories are insufficient to explain this remarkable advancement. With respect to educational levels, for instance, it is not immediately clear why Southern Europeans (who immigrate with the lowest level of education) advance economically more quickly than Western Europeans, who immigrate with relatively high levels of education. Moreover, human capital theories do not explain why immigrants of sub-Saharan African origin (who have higher levels of education than Southern Europeans) remain economically inferior to immigrants of both Southern and Western European backgrounds, and are unable to compensate for their ethnic penalty.

Other models have explained these discrepancies by focusing on different mechanisms, such as those related to cultural barriers and economic integration (see Heath, Rothon, and Kilpi 2008). In addition, differences have been explained by an unobserved lack of readiness for the U.S. labor market as well as difficulties in transferring skills from the origin countries (e.g., Chiswick 1978). While these models seem reasonable when comparing minorities in the United States who were born in different countries, they do not fully explain the differences between different minority groups born in Western Europe. To explain the contrasting patterns of economic advancement and immigration, we need more quantitative and qualitative examinations of the immigrant groups we investigated here. In addition, a comparison, over time, of non–Western Europeans who stay in Europe with those who immigrated to the United States can shed light on additional mechanisms at play.

Indeed, the changing demographics of Western European immigration to the United States have important implications. The "white migration" that formed the cornerstone of U.S. society throughout the twentieth century is becoming increasingly "colorful," especially considering the relatively high rates of return of Western Europe majority groups following immigration to the United States. As a result, the nature of human capital provided by immigration from Western Europe is shifting, and the labor market in the United States may be faced with changes in workers' skills and gender divisions of labor. There may also be consequences for the U.S. education system and school curricula, which will need to adapt to incorporate children and families from non-Western cultures. In particular, the once Christian-dominated immigration from Western Europe has become increasingly religiously diverse. Adjustments to the education system are of particular importance in the states that receive a larger number of non-Western and religious minority migrants (see Figure 2).

Finally, it is important to note that race, ethnicity, and place of birth no longer necessarily coincide with one another. As a result, policies that treat groups of immigrants, such as those from Western Europe, as largely homogeneous may not accurately reflect the increasingly diverse and complex reality. Indeed, it seems that many of Western Europe's minorities may be demographically and socially similar to potential immigrants from the Eastern hemisphere. Accordingly, the immigration quota system in the United States that continues to differentiate between immigrants from the Eastern and Western hemispheres should be reexamined.

References

Aeberhardt, Romain, Denis Fougère, Julien Pouget, and Roland Rathelot. 2010. Wages and employment of French workers with African origin. *Journal of Population Economics* 23 (3): 881–905.

Aigner, Dennis, and Glen Cain. 1977. Statistical theories of discrimination in labor markets. *Industrial and Labor Relations Review* 30 (2): 175–87.

Alba, Richard, and Victor Nee. 1997. Rethinking assimilation theory for a new era of immigration. *International Migration Review* 31 (4): 826–74.

Alba, Richard, and Victor Nee. 2005. *Remaking the American mainstream: Assimilation and contemporary immigration*. Cambridge, MA: Harvard University Press.

Becker, Gary S. 1971. *The economics of discrimination*. Chicago, IL: University of Chicago Press.

Borgna, Camilla, and Dalit Contini. 2014. Migrant achievement penalties in Western Europe: Do educational systems matter? *European Sociological Review* 30 (5): 670–83.

Borjas, George. 1982. The earnings of male Hispanic immigrants in the United States. *Industrial and Labor Relations Review* 35 (3): 343–53.

Borjas, George. 1987. Self-selection and the earnings of immigrants. *The American Economic Review* 77 (4): 531–53.

Borjas, George. 2001. *Heaven's door: Immigration policy and the American economy*. Princeton, NJ: Princeton University Press.

Castles, Stephen, Hein de Haas, and Mark J. Miller. 2013. *The age of migration: International population movements in the modern world*. New York, NY: Guilford Press

Ceobanu, Alin M., and Xavier Escandell. 2010. Comparative analyses of public attitudes toward immigrants and immigration using multinational survey data: A review of theories and research. *Annual Review of Sociology* 36:309–28.

Chiswick, Barry. 1978. The effect of Americanization on the earnings of foreign-born men. *The Journal of Political Economy* 86 (5): 897–921.

CIA. 2013. Distribution of Family Income - Gini index. Washington, DC: CIA.

Cohen, Yinon. 1996. Economic assimilation in the United States of Arab and Jewish immigrants from Israel and the Territories. *Israel Studies* 1 (2): 75–97.

Cohen, Yinon, and Yitzchak Haberfeld. 1997. The number of Israeli immigrants in the United States in 1990. *Demography* 34 (2): 199–212.

Cohen, Yinon, and Andrea Tyree. 1994. Palestinian and Jewish Israeli-born immigrants in the United States. *International Migration Review* 28 (2): 243–55.

Coleman, David, Paul Compton, and John Salt. 2002. Demography of migrant populations: The case of the United Kingdom. In *The demographic characteristics of immigrant populations. Population Studies*, eds., Werner Haug, Paul Compton, and Youssef Courbage, 497–552. Strasbourg, France: Council of Europe Publishing.

Constant, Amelie F., Martin Kahanec, and Klaus F. Zimmermann. 2009. Attitudes towards immigrants, other integration barriers, and their veracity. *International Journal of Manpower* 30 (1/2): 5–14.

Crul, Maurice, and Hans Vermeulen. 2003. The second generation in Europe. *International Migration Review* 37 (4): 965–86.

Fossett, Mark. 2006. Ethnic preferences, social distance dynamics, and residential segregation: Theoretical explorations using simulation analysis *Journal of Mathematical Sociology* 30 (3/4): 185–273.

Freeman, Gary P. 2004. Immigrant incorporation in Western democracies. *International Migration Review* 38 (3): 945–69.

Gans, Herbert J. 2007. Acculturation, assimilation and mobility. *Ethnic and Racial Studies* 30 (1): 152–64.

Garssen, Joop, and Han Nicolaas. 2008. Fertility of Turkish and Moroccan women in the Netherlands: Adjustment to native level within one generation. *Demographic Research* 19:1249–79.

Heath, Anthony. 2007. Crossnational patterns and processes of ethnic disadvantage. In *Unequal chances: Ethnic minorities in Western labour markets*, eds. Anthony Heath and Sin Yi Cheung, 639–95. Oxford: British Academy, Oxford University Press.

Heath, Anthony, and Sin Y. Cheung. 2006. *Ethnic penalties in the labour market: Employers and discrimination*. Leeds: Department for Work and Pensions.

Heath, Antony, and Sin Yen Cheung, eds. 2007. *Unequal chances: Ethnic minorities in Western labour markets*. Oxford: Oxford University Press.

Heath, Anthony, and John Ridge. 1983. Social mobility of ethnic minorities. *Journal of Biosocial Science* 15 (S8): 169–84.

Heath, Anthony, Catherine Rothon, and Elina Kilpi. 2008. The second generation in Western Europe: Education, unemployment, and occupational attainment. *Annual Review of Sociology* 34:211–35.

ISSP Research Group. 2013. National Identity 3. Cologne, Germany: GESIS, International Social Survey Programme.

Johnson, Dawn R., Matthew Soldner, Jeannie Brown Leonard, Patty Alvarez, Karen Kurotsuchi Inkelas, Heather T. Rowan-Kenyon, and Susan D. Longerbeam. 2007. Examining sense of belonging among first-year undergraduates from different racial/ethnic groups. *Journal of College Student Development* 48 (5): 525–42.

Kislev, Elyakim. 2014. The effect of minority/majority origins on immigrants' integration. *Social Forces* 92 (4): 1457–86.

Kislev, Elyakim. 2015. The transnational effect of multicultural policies on immigrants' identification: The case of the Israeli diaspora in the USA. *Global Networks* 15 (1): 118–39.

Kogan, Irena. 2006. Labor markets and economic incorporation among recent immigrants in Europe. *Social Forces* 85 (2): 697–721.

Lieberson, Stanely. 1985. Unhyphenated whites in the United States. *Ethnic and Racial Studies* 8 (1): 159–80.

Lieberson, Stanely, and Mary Waters. 1985. Ethnic mixtures in the United States. *Sociology and Social Research* 70:43–52.

Neidert, Lisa J., and Reynolds Farley. 1985. Assimilation in the United States: An analysis of ethnic and generation differences in status and achievement. *American Sociological Review* 50 (6): 840–50.

Nielsen, Helena S., Michael Rosholm, Nina Smith, and Lief Husted. 2003. The school-to-work transition of second generation immigrants in Denmark. *Journal of Population Economics* 16 (4): 755–86.

Parrillo, Vincent, and Christopher Donoghue. 2005. Updating the Bogardus social distance studies: A new national survey. *The Social Science Journal* 42 (2): 257–71.

Polyakova, Alina. 2015. *The dark side of European integration: Social foundations and cultural determinants of the rise of radical right movements in contemporary Europe*. vol. 4. New York, NY: Columbia University Press.

Röder, Antje, and Peter Mühlau. 2011. Discrimination, exclusion and immigrants' confidence in public institutions in Europe. *European Societies* 13 (4): 535–57.

Ruggles, Steven, Trent Alexander, Katie Genadek, Ronald Goeken, Matthew Schroeder, and Matthew Sobek. 2013. Integrated Public Use Microdata Series: Version 5.0 [Machine-Readable Database]. Minneapolis, MN: Minnesota Population Center.

Safi, Mirna. 2010. Immigrants' life satisfaction in Europe: Between assimilation and discrimination. *European Sociological Review* 26 (2): 159–76.

Semyonov, Moshe, Rebeca Raijman, and Anastasia Gorodzeisky. 2006. The rise of anti-foreigner sentiment in European societies, 1988–2000. *American Sociological Review* 71 (3): 426–49.

UN. 2013. Human Development Index (HDI). New York, NY: Human Development Report Office, the United Nation. Available from http://hdr.undp.org/en/statistics/.

Van Tubergen, Frank, Ineke Maas, and Henk Flap. 2004. The economic incorporation of immigrants in 18 Western societies: Origin, destination, and community effects. *American Sociological Review* 69 (5): 704–27.

Wodak, Ruth, and Salomi Boukala. 2015. European identities and the revival of nationalism in the European Union: A discourse historical approach. *Journal of Language and Politics* 14 (1): 87–109.

World Bank. 2013. Gini index. Washington, DC: World Bank.

Zhou, Min. 1997. Segmented assimilation: Issues, controversies, and recent research on the new second generation. *International Migration Review* 31 (4): 975–1008.

Quilting a Time-Place Mosaic: Concluding Remarks

By
BARBARA ENTWISLE,
SANDRA L. HOFFERTH,
and
EMILIO F. MORAN

Social science is at a pivotal moment. The advent of "big data" from the Internet, social media, and smartphones as well as newly available administrative data from electronic sources has opened the door to new understandings of people and society. That said, realizing this promise requires a vision for the future and a practical plan for reaching it. The articles in this volume begin this work. Each addresses some aspect of data linkage. Each can be considered a patch in a time-place mosaic. This concluding article considers the articles as a collection and how they might be quilted together. It discusses the diversity of sources available, differences in time depth and sociospatial coverage, and the many challenges of using data not designed for research. It identifies the benefits that would cumulate as a set of regional data centers to assemble, link, curate, and share diverse data sources is established and coordinated. These data centers could provide the combination of national coverage, local depth, and temporal precision needed to advance our understanding of the American population.

Keywords: big data; social science; data linkage; data access; confidentiality; survey research; administrative data; social media

The social and behavioral sciences are at a remarkable juncture. Diverse data from transactional sources, the Internet, social media, and other sources not designed for

Barbara Entwisle is Kenan Distinguished Professor of Sociology and a fellow of the Carolina Population Center at the University of North Carolina at Chapel Hill. Her research focuses broadly on the study of social, natural, and built environments and their consequences for a range of demographic and health outcomes.

Sandra Hofferth is professor emerita, School of Public Health, and a research professor, Maryland Population Research Center, University of Maryland. Her research focuses on American's use of time; economic disadvantage, parental behavior, and child health and development; fathers and fathering; and immigrant youth's transition to adulthood.

Correspondence: entwisle@unc.edu

DOI: 10.1177/0002716216683698

research can now be used in research to varying degrees, making it possible to ask and answer questions about the dynamics of social interaction and behavior in new ways. This is particularly true when these data are combined with social surveys and other traditional social science data sources. Currently, however, the promise of these new opportunities is unfulfilled. One reason is that individual researchers and teams solve challenges associated with the use of new data alone, independently, for their own purposes. There is a potential for duplication of effort. Further, the linkages these researchers create, and the datasets they build, are not integrated into a larger, comprehensive resource. The vision of a social observatory or set of regional data centers laid out in the first article and that led to this volume is to build on and coordinate these efforts, strategically leveraging them to benefit social, behavioral, and economic science, as well to create a public good for the nation.

Our vision is to create a national resource that would assemble, document, curate, and disseminate the growing collection of data potentially relevant to social, behavioral, and economic research—a social observatory. Such a resource would make it possible to ask new questions and to answer others in new ways. It would promote data sharing, reduce duplication, and increase the cost-effectiveness of research. It would also enhance research quality and the reproducibility of results. There are many ways in which such a resource might be organized. Although it is possible to think in terms of a single national social observatory, it is more likely that there would be multiple entities, regional or distinguished according to foci that would cluster expertise (e.g., social programs, social mobility, migration, health, environment-related hazards). What is critical, however, is that these centers or observatories would play a key role in linking data and making them available to the rest of the science community and society. The challenges of linking data are significant and cannot be downplayed; it is for this reason that this volume—and the network of data centers it proposes—sees as a priority advancing the state of the science on this subject

The articles assembled for this special issue all address some aspect of data linkage. They cover a broad range of topics, including monitoring the health and well-being of the population (Dai et al.; Bader et al.); assessing the impact of social programs (O'Hara et al.; Digitale et al.; Leonard et al.); measuring concepts such as family and neighborhood in new ways (Bader et al.; O'Hara et al.; Browning et al.); planning for and responding to environmental hazards (Fussell et al.); and developing new approaches to movement, mobility, and migration (Browning et al.; Kislev; Fussell et al.). Linking diverse data sources is central to all of them.

In these concluding remarks, we consider the articles as a collection. We comment on the breadth of data sources used here, ways that data can be linked, and

---

Emilio F. Moran is Hannah Distinguished Professor at Michigan State University. He is the author of eleven books, fifteen edited volumes, and more than 200 journal articles and book chapters. His research addresses how humans interact with the environment. He was elected to the U.S. National Academy of Sciences in 2010.

NOTE: Funding for the Social Observatories Coordinating Network was provided by National Science Foundation grant SES1237498.

potential benefits to research quality that might flow from linking data. We also discuss challenges associated with linking diverse data in ways that are accurate and cumulative and with making the linked data accessible to a broad range of researchers. Collectively, the articles demonstrate the value of a social observatory that assembles sources of data on multiple topics, multiple types of information on a single topic, and even multiple measures of the same concept within a topic (e.g., measures of the family).

Sources, Links, and Coverage

The articles in this volume illustrate the breadth and diversity of data relevant to social, behavioral, and economic science, and in so doing reflect much of the opportunity as well as many of the challenges associated with their use. Included in these kinds of data are census, surveys, and traditional sources of administrative data. Also included are less traditional sources of administrative data such as from the Internal Revenue Service (IRS), the Department of Housing and Urban Development, and other federal agencies; state-level program data from the Temporary Assistance for Needy Families (TANF) program, Supplemental Nutrition Assistance Program (SNAP), and the Women, Infants, and Children (WIC) nutrition program; and data from nonprofits such as the Crossroads Program in Dallas, Texas. Data from "new" sources such as electronic medical records, Internet data (e.g., Google StreetView), social media (e.g., Twitter), and locational data from GPS-enabled mobile devices such as smartphones are also considered.

Linking these data is accomplished in a variety of ways. Sometimes, linking is at the person level, as discussed by O'Hara and her colleagues in their article. This requires careful attention to the privacy of the individuals involved and the confidentiality of their responses and, ultimately, may necessitate strict limits on access to the linked data. The datasets described by O'Hara et al. reside in highly secure data enclaves managed by the U.S. Census Bureau. Potential users of the data must go through a rigorous application and review process, and if approved, use the data onsite with strict oversight of any tables or graphs created. Privacy and confidentiality concerns can also arise when linking at the household level. Rather than limiting access, Leonard and her colleagues anonymize their data as a way to address these concerns while making the data more broadly available. Digitale and colleagues link individuals with the family-planning facilities they reported using; to maintain confidentiality, only broad characteristics of the facilities are analyzed. Bader and his team point out that identifying individual respondents' addresses in proprietary software may violate confidentiality and limit their use in conducting neighborhood audits.

Typically, linking is accomplished at a higher geographic level such as a block group (Browning et al.), census tract (Bader et al.), county (Fussell et al.), state (Dai et al.), or country (Kislev). It is worth noting that in many instances, less aggregated data are available. For example, Browning and his colleagues code

CONCLUSION 193

latitude-longitude coordinates collected by GPS-enabled smartphones to block groups rather than analyze them directly. In the most finely disaggregated analysis, Digitale and colleagues link participant residence with that of family-planning facilities within 10 kilometers using GPS coordinates. Dai and colleagues link Twitter data to state-level prevalence of asthma and other state characteristics from national surveys. It is theoretically possible for them to link data at lower levels of aggregation; however, at lower levels the precision of estimates of asthma prevalence is likely to be poor. Data at higher levels of aggregation are less likely to involve privacy and confidentiality concerns, although these cannot be entirely ruled out.

Linking data also involves temporal considerations. To be useful, linked data sources need to be temporally aligned, with putative causes measured at or before effects. Some of the new sources make data available instantaneously and continuously, in contrast to the annual, biannual, quinquennial, or decadal time steps of traditional sources. In the collection of articles that we have assembled, the finest temporal unit of analysis is the day, as presented by Dai and colleagues in their analysis of Twitter data linked to the Behavioral Risk Factor Surveillance System (BRFSS) and the American Community Survey (ACS). Temporally speaking, although fine resolution is available in the data sources, Dai and colleagues yield on geographic scale, grouping data into states or grouping states by quartiles or population characteristics. The same tendency toward geographic aggregation is found among others working at fine temporal scale. Indeed, there tends to be an inverse relationship between temporal and geographic detail. This relationship may reflect the capacity of individual researchers and the size of datasets that are easily managed. With proper support, regional data centers would not be limited in this way. These tradeoffs reflect the limitations of many individual research projects in the social and behavioral sciences, which are limited by small awards, small teams, and the need to reduce time and spatial coverage to achieve sufficient depth. Observatories, as long-term data centers, would go a long way toward helping to address these limitations.

Spatial and temporal coverage varies in this volume. Some of the authors assemble data that are national in scope. For example, Fussell and her colleagues integrate 1970 to 2010 data from the Spatial Hazard Events and Losses Database for the United States (SHELDUS) and U.S. census data using ArcGIS geo-referenced county-year FIPS codes and county-boundary files. Dai et al.'s and Kislev's studies are also national in scope but not as expansive in temporal coverage. Other authors focus on the local level. For example, Bader and his colleagues examine aging in place in four cities: New York, San Jose, Philadelphia, and Detroit. Browning and his colleagues examine adolescent exposure to violent locations in Columbus, Ohio. Leonard, Hughes, and Pruitt investigate responses of vulnerable households to unanticipated health events such as heart attacks, which they label "health shocks," in Dallas, Texas. It would be desirable to scale up these studies, or at least the data on which they are based, to a national level so that social and geographic inequalities and the typicality of these cases could be better understood. It would also be desirable to extend them temporally.

Finally, although we talk about social observatories in terms of research in the United States, it may make sense not to restrict the focus in this way. We live in a globalized world in which the flow of people does not stop at national boundaries, and our studies must follow people to understand their behavior. An example of this here is Kislev's article, which draws on survey data from Europe as well as census and survey data from the United States to examine immigrant outcomes in the United States. Although Dai et al. focus on the United States, they have access to Twitter data from countries around the world. Digitale and her colleagues link survey and administrative data in Malawi to look at the impact of family-planning facilities on contraceptive use among young women residing there. These articles remind us that social science is global; important questions are not confined to national borders. Indeed, some of the newer sources of data are not contained within nation-states. This is particularly true for social media data, where geographic boundaries are in a very real sense arbitrary.

Improved Measurement and Inference

Collectively, the articles in this volume demonstrate the value of a network of observatories or data centers for addressing topics critical to the advancement of social and behavioral science. In the process, they comment on measurement issues and potential problems of statistical inference that are worthy of consideration. We review the value of linked data for addressing these more methodological topics here.

One value of linked sources is that each can provide a check on the others. This is especially useful in leveraging new sources of data. The new "big data" sources have important advantages such as temporal and spatial detail more consistent with a dynamic and place-based understanding of social patterns and behavior than typically possible with traditional sources. However, because they were designed for purposes other than research, there are important questions to be asked about their provenance, availability, and quality, including coverage, representativeness, completeness, accuracy, and relevance of the information provided. For example, a moment-by-moment picture of some aspects of behavior can be obtained from the Internet, Twitter, and mobile devices, in ways that are not possible with traditional data sources. Whether this picture is representative is an open question. Not all Twitter data can be linked to a geographic unit, even a country—a problem noted by Dai and colleagues. Furthermore, data may not be population-representative; Twitter is used more by men than women, more by younger than older adults. Although traditional sources such as social surveys collect data only annually or at greater intervals, they are carefully designed and their coverage is well-documented. Hao, Lee, and Dai capitalize on contrasting strengths when they use BRFSS data to evaluate the potential use of Twitter data for monitoring asthma at the state level. They conclude that tweets can serve "as a rapid, cost-effective health detection system with real-time information to monitor chronic disease and track public sentiment," although correlations of asthma-related tweets with BRFSS reports of asthma prevalence at the state level are modest (r = .32 to .38).

Another strength of nontraditional sources is potentially better coverage of hard-to-reach and frequently underreported groups. For example, Leonard and her colleagues argue that, in Dallas, vulnerable populations are better captured through visits to the Crossroads Food Pantry and electronic medical records from a local hospital than through traditional data sources. Further, while each of these sources may have limitations, together they may provide a more comprehensive picture than any of them could alone. Along similar lines, O'Hara and her colleagues report on studies examining administrative records on participants in state-supported social programs such as WIC, TANF, and SNAP and compare them to the self-reported participation from the ACS and the Survey of Income and Program Participation. These studies find underreporting of program participation ranging from 10 percent to 35 percent. Yet, turning it around, they report that supplementing census records with WIC and SNAP records substantially improves the coverage of young children, for example. The complementary perspectives of multiple sources may provide a comprehensive picture of the phenomenon in question. Centralizing the assembly of these data makes it possible to play to the strengths of one against the strengths of the other—making the challenge of linking data all the more necessary and valuable.

Information available in one dataset may be used to supplement another, as a way to address unit and item nonresponse. Plans for the 2020 U.S. Census call for the use of administrative data as a supplement to the main data collection (O'Hara et al.). In addition to reducing missing data, using administrative datasets to "pre-fill" already known information into survey instruments reduces the burden in the interview.

Bader and his colleagues comment on how valuable multiple data sources are for addressing potential "same source" bias. When measures for both hypothesized causes and effects come from the same source, distinctive features of that source may lead to correlated error, which in turn may lead to biased estimates of causal effects in statistical analysis. When multiple sources are available, this problem can be avoided by substituting one source for the other. Multiple sources also make it possible to use sophisticated approaches to measurement such as multitrait multimethod and other types of structural equation models. A multitrait multimethod model is one type of model.

Finally, linked data make possible entirely new approaches to examining key social science concepts. O'Hara and her colleagues capitalize on linked data to explore family units. In their definition, families are sets of relationships that shift over time as unions are formed and dissolved, children are born, and family members move and die. Families are related to, but not synonymous with, households. It takes multiple datasets to identify family relationships and track them over time. In another example, Browning and his colleagues show that neighborhoods need not be identical for everyone who lives in a particular location. Other geographies may come into play. With GPS-enabled smartphones, it is possible to take an individualized approach to daily activity patterns and, in the process, redefine what is meant by "neighborhood."

Challenges

As the articles demonstrate, linking data from multiple sources provides many benefits. There are also challenges—How do we link diverse forms of data in a way that is accurate, cumulative, and accessible to a broad range of researchers?

As the articles show, the quality of the link between data sources is critical, as is clear documentation of any problems encountered, particularly when newer, nontraditional sources are used. For example, location information needed to link data may be missing or of questionable quality. Dai and colleagues impute geographic location of tweeters based on their self-reported city, state, and country location in the Twitter metadata. Of all the tweets in the 1 percent public access sample they were using, approximately 25 percent had country information, and of these, 16 percent were from the United States. In interpreting this statistic, it would be helpful to know whether tweeters in the United States were more or less likely to reveal their country of residence than tweeters from other countries. Whatever the percentage of U.S. tweeters who report that they are based in the United States, it will fall short of 100 percent, perhaps substantially so, and this will degrade links to other data sources. Fortunately, of those identifying as a U.S. tweeter, almost all (91 percent) supplied a state of residence, a much more reassuring statistic. Even when links exist, there may be questions about their quality. First, the choice of keywords can have a huge impact on the correlation and prediction results. How should researchers choose keywords? Is it based on review of the literature, directly extracted from other sources, or from domain experts? It is important to select and justify these keywords carefully. Second, the data are noisy. In the Dai et al. study, for example, nearly 6 percent of asthma-related tweets were from the top ten users and 19 percent had URLs, suggesting that some of these tweets might have been sent through twitter bots or spammers. Third, assessing sentiment from the language in tweets is not an exact science. Given the complexity of language, true emotions or feelings might not be captured.

An observatory or set of regional data centers would be an ideal place to develop such information and make it broadly available to users. Variables would be created and links performed and documented cumulatively. Indeed, doing so creates economies of scale. Many researchers are creating the same measures—tract-level measures of poverty, race-ethnic composition, and immigrant status—from the same sources. Making these measures available centrally would save time and money. It could also enhance the reproducibility of results, as common variables would be carefully constructed and assessed by experts.

Data could be assembled once, rather than separately by each project. For instance, Bader and his colleagues use Google Street View as a less costly alternative to neighborhood audits, but even this approach can be expensive. To keep costs down, street segments were sampled, only one side of the street was coded, interpolation techniques were used to generate a spatially continuous surface, and only four cities were analyzed. With a coordinated and centralized approach, it might be possible to include all of the street segments, not only a sample, for all of the country, not only four cities. That way, observationally based measures such as neighborhood disorder could be developed for variably defined spatial units (including custom definitions) and then used by many researchers, for many purposes. For example, one can imagine that neighborhood disorder might be useful in Browning et al.'s assessment of adolescent exposure to violence. With automation, it is becoming increasingly feasible to think of national data coded at the street segment level. Whether variables developed from Street View could be made available in this way would depend on arrangements with Google and their willingness to allow it.

Special arrangements would be needed for other sources of data as well. For example, Fussell and her colleagues purchased the Spatial Hazard Events and Losses for the United States database from the University of South Carolina for their analysis of environmental hazards and migration. Costs are modest for academic users, but subscriptions for governmental, nonprofit, and especially corporate use are more expensive. The license agreement makes it clear that the data may not be shared further.[1] In cases such as this one, the observatory might provide easily applied protocols for linking to these data rather than the data themselves.

Access is central to our concept of a social observatory. The U.S. Census Bureau houses data not only from its data collections but also from federal agencies including the IRS, the Department of Housing and Urban Development, the Social Security Administration, and the Centers for Medicare and Medicaid Services, as well as state-level data on TANF, SNAP, and WIC. As O'Hara and her colleagues show, when combined, these data sources open up new approaches to the study of families and their complexities. However, access to these resources is limited to projects that benefit the Census Bureau and are conducted within their highly secure Remote Data Centers. We imagine a complementary resource in which data are more readily available. Well-designed and thoroughly tested approaches to sharing even highly confidential data are available that allow for a broader array of potential uses (e.g., see those developed by the National Longitudinal Study of Adolescent to Adult Health).[2]

Wrapping Up

Each article in this special issue is a valuable contribution in its own right. Each discusses the data sources and linking challenges associated with a particular social or behavioral science question of interest to the authors. As we discussed here, the articles are also valuable as a collection. They demonstrate a variety of approaches to linking diverse datasets, address the tradeoffs of temporal and spatial detail, raise important questions about access, illustrate a variety of challenges associated with the use of data not designed for research, and set an agenda for the future. Additionally, as a collection, they show many different ways in which linking data from diverse sources can improve analyses beyond the research possible with only one of those sources. They show how diverse sources with varying strengths might be leveraged and how linking them can improve research that is done with only one of those sources. New questions can be addressed. We know that social mobility and life chances vary from place to place in the United States; but why? What can be done to reverse the erosion of the middle class? Improved identification of vulnerable populations is critical if we are to develop more compassionate public and private social welfare programs to serve these populations. Concepts such as family and neighborhood can be defined in innovative ways with new data sources. Indeed, family arrangements have changed dramatically over the past half century, and yet our ways of measuring and describing them have not.

Collectively, the articles offer broad commentary on the potential value of and challenges associated with the creation of a social observatory, especially the linked data that would be central to such a resource, and point to next steps. We can imagine using them and especially the data on which they are based, to quilt a time-place mosaic and have an efficient and national resource for improving science, policymaking, and services for the benefit of the American people. In broad strokes, we can paint a picture of what this resource might look like in the longer term. Consistent with federal policy, it will promote data sharing and re-use. In this way, it follows in the footsteps of the infrastructure social surveys that have served the needs of the social sciences so admirably for the past half century, including the National Longitudinal Surveys, the Panel Study of Income Dynamics, the American National Election Studies, and the General Social Survey (GSS). As one example, take the GSS, which is funded by the National Science Foundation. Its director estimates that there have been more than 27,000 articles, dissertations, books, and conference papers based on the GSS, and that each year, 400,000 students use the GSS in a class they take. In 2016, GSS data informed news coverage of such diverse topics as racial attitudes, child care, corporal punishment, and marital happiness. This model contrasts with the situation as it exists now with the new "big data." Many research projects using these data develop their linkages and measures independently, for their own purposes. In some cases, the same variables based on the same source are created again and again, a duplication of effort that we can ill afford. One such example is neighborhood poverty rates based on the ACS, a product of the U.S. Census Bureau that has been used by hundreds of studies. A centralized source for such measures would create efficiencies, and also, because of improved oversight, contribute to robust and rigorous social science. Each study represents one patch in the time-place mosaic referenced above. If researchers contributed the measures they create—i.e., other patches—based on some of the newer data sources such as Twitter, Google Street View, and administrative records, this could be of broad benefit to social science and society. The local variability in social mobility and life chances documented in the introduction to this volume, and on full display in the recent election, would be fully captured. When fully assembled, the set of observatories or data centers will provide the combination of national coverage, local depth, and temporal precision needed to advance our understanding of the American population.

Notes

1. http://hvri.geog.sc.edu/SHELDUS/docs/END_USER_LICENSE_AGREEMENT.pdf
2. http://www.cpc.unc.edu/projects/addhealth/contracts/add-health-contracts-homepage.

SAGE Deep Backfile Package

Content ownership is becoming increasingly important in hard budgetary times. Investing in the SAGE Deep Backfile Package means owning access to over 400 SAGE journal backfiles.

5 good reasons to own the deep archive from SAGE...

1. Breadth
SAGE has collected over 400 journal backfiles, including over 500,000 articles of historical content covering interdisciplinary subjects in business, humanities, socials science, and science, technology and medicine.

2. Depth
SAGE's deep backfile coverage goes to volume 1, issue 1; through the last issue of 1998 (content from January 1999 to the present is included in the current subscription). You will own content spanning over a century of research. Our oldest article is from 1879 in **Perspectives in Public Health** (formerly *The Journal of the Royal Society for the Promotion of Health*).

3. Quality
We pride ourselves on high-quality content, meeting our markets' need for interdisciplinary, peer-reviewed, journal backfiles to provide your library. Close to 50% of the journals in the entire **SAGE Deep Backfile Package** are ranked in the Thomson Reuters Journal Citation Reports®.

4. Award-winning *SAGE Journals* online delivery platform
Materials are easy to find on *SAGE Journals* (SJ), hosted on the prestigious HighWire Press platform.

5. Pricing
We offer **flexible backfile purchase and lease options** to accommodate library budgets of all sizes. This package option offers the most value for your money, including great savings off list price for individual journal backfile purchases.

Need something more specific?

Titles included in the **SAGE Deep Backfile Package** are also available in smaller, discipline-specific packages:

- **Humanities and Social Science (HSS) Backfile Package**
- Scientific, Technical, and Medical (STM) Backfile Package
- Health Sciences Backfile Package
- Clinical Medicine Backfile Package

For more information, contact
librarysales@sagepub.com

BRADNER LIBRARY
SCHOOLCRAFT COLLEGE
LIVONIA, MICHIGAN 48152

ⓢSAGE track

Authors!
Submit your article online with SAGE Track

SAGE Track is a web-based peer review and submission system powered by ScholarOne® Manuscripts

The entire process, from article submission to acceptance for publication is now handled online by the SAGE Track web site. 300 of our journals are now on SAGE Track, which has a graphical interface that will guide you through a simple and speedy submission with step-by-step prompts.

SAGE Track makes it easy to:

- Submit your articles online

- Submit revisions and resubmissions through automatic linking

- Track the progress of your article online

- Publish your research faster

To submit a manuscript, please visit:
http://www.sagepub.com/journalsIndex.nav
Select a journal and click on the Manuscript Submissions tab for detailed instructions for submission.